Muted Voices

Muted Voices

The Recovery of Democracy in the Shaping of Technology

Jesse S. Tatum

Lehigh
University
Press

Bethlehem: Lehigh University Press
London: Associated University Presses

Associated University Presses
440 Forsgate Drive
Cranbury, NJ 08512

Associated University Presses
16 Barter Street
London WC1A 2AH, England

Associated University Presses
P.O. Box 338, Port Credit
Mississauga, Ontario
Canada L5G 4L8

The paper used in this publication meets the requirements of the American National Standard for Permanence of Paper for Printed Library Materials Z39.48-1984.

Library of Congress Cataloging-in-Publication Data

Tatum, Jesse S., 1952–
    Muted voices : the recovery of democracy in the shaping of
technology / Jesse S. Tatum.
        p.   cm.
    Includes index.
    ISBN 0-934223-58-0 (alk. paper)
    1. Technology—Social aspects.   2. Democracy.   I. Title.
T14.5.T37   2000
303.48′3—dc21   99-32780
                                                               CIP

Contents

Acknowledgments

I OWE SPECIAL THANKS TO THE HUNDREDS OF INDIVIDUAL CITIZENS who have freely shared many hours of their time with me as I have worked on the two studies presented in this book. My own efforts have been sustained primarily by their continuing sense of commitment to a better world, and it is my sincere hope that what I have written here has captured something of the substantive message they might wish me to convey, as well as some portion of their positive spirit.

I am greatly indebted, as well, to the National Science Foundation, the U.S. taxpayer, and Dr. Rachelle Hollander of NSF's Ethics and Values Studies Program for their support of the home power study reported in Part II.

Thanks are also due to the faculty of the Department of Science and Technology Studies at Rensselaer Polytechnic Institute, who so warmly welcomed me as a Visiting Fellow and Visiting Research Scholar in the period 1994 to 1996. Ned Woodhouse's patient and incisive comments on more than one occasion, and other substantive and logistical contributions from Langdon Winner, Steve Breyman, John Schumacher, and Ed Hackett, have all been greatly appreciated.

Finally, I would like to thank Steve Cutcliffe, Carl Mitcham, and Ned Woodhouse for hosting interim presentations and contributing to very useful discussions at their respective institutions during particularly critical stages of my work. And I would like to thank all of those listed above, along with Albert Borgmann, David Strong, David Noble, Brian Martin, Brian Rappert, David Levinger, Todd Cherkasky, John Monberg, and the many other STS scholars of substance whose integrity and continuing efforts to respond to real and pressing human needs remain a source of inspiration and encouragement.

In this august company, the reader may now expect more than

I or this book can begin to deliver. Any apologies must be mine as neither the National Science Foundation nor any of those named (or unnamed) above share responsibility for any errors, oversights, misapprehensions, or other misconstructions the reader may encounter in the pages that follow.

* * *

The following publishers have generously given permission to use quotations from copyrighted works. Lewis Mumford. Authoritarian and Democratic Technics. *Technology and Culture,* vol. 5, no. 1, pp. 1–9, © 1964, Society for the History of Technology. Albert H. Wurth, "Public Participation in Technological Decisions: A New Model" *The Bulletin of Science, Technology, and Society,* vol. 12, pp. 289–293, © 1992 by Sage Publications, Inc. Reprinted by Permission of Sage Publications, Inc. Andrew D. Zimmerman, "Toward a More Democratic Ethic of Technological Governance" *Science, Technology & Human Values,* vol. 20, no. 1, pp. 86–103, © 1995 by Sage Publications, Inc. Reprinted by Permission of Sage Publications, Inc. Benjamin Barber, *Strong Democracy: Participatory Politics for a New Age.* © 1984 The Regents of the University of California, Reprinted by permission, University of California Press, Berkeley, California. David Rose, "Continuity and Change: Thinking in New Ways About Large and Persistent Problems" *Technology Review,* Feb/March 1981. Reprinted with permission from *Technology Review,* published by the Association of Alumni and Alumnae of MIT, copyright 1999.

Introduction

THIS BOOK IS MOTIVATED BY CONCERNS ABOUT THE SOCIAL, POLITI-
cal, and material implications of the nonparticipatory nature of
most technology design and development in the United States.
Citizen participation that is neither direct nor vigorous must,
after all, give us reason to worry that the legitimacy of technology's advance may fall ever more disturbingly into question as the
prospects for crippling dissent rise in opposing proportions. Ultimately, participation is essential to technology's responsiveness
to the interests and value commitments of citizens. Cut adrift
from such ordinary interests, technology design eventually risks
material misdirection and even the progressive subversion of material sustainability, as an era of environmental awakening
should by now have taught us.

Strategies for easing these hazards will be sought here in exceptions to the nonparticipation rule. In particular, two detailed
ethnographic studies will be offered, describing uniquely vigorous
citizen involvement in efforts to shape technology in the pivotally
important field of energy production and use. The first study of
what might loosely be termed a grass-roots antinuclear group
from the 1980s unexpectedly reveals deep positive commitments
to a reformulation of community and human/nature relationships that have long been frustrated by the more conventional
value commitments structured into the shaping of modern technology. The second study, which moves into new terrain beyond
any reaction to authoritatively advanced technologies, describes
the emergence from the margins of society of a vigorous set of
positive technological alternatives centered around renewable
(photovoltaic, microhydro, and small wind) "home power" systems. The home power movement, involving more than one hundred thousand homes as of 1993, reveals commitments similar to
those of the antinuclear movement but also the surprising fruits
of an elaborate and extended participatory research effort that is

9

beginning to achieve the expression of new community and human/nature relationships through new technological configurations. Home power developments offer dramatic confirmation of the possibility of citizen-centered and -initiated technology design and development, as well as the possibility of coherent and attractive expressions of alternative perceptions and value commitments through technology.

Richly detailed presentations of both studies offer new insights into the ways in which technology forms our lives. More importantly, however, the two studies provide vivid indications of the need to listen more deeply for the muted voices in our society whose sometimes inchoate message remains essential to a *democratic* shaping of technology. The book ends with the outlines of a promising new model for the recovery of democracy in the shaping of technology and with some discussion of its possible implications.

BACKGROUND

Fundamental to the frame of thought adopted in this book is the notion that technology is "underdetermined"[1] by science. That is, the technology we take for granted in our daily lives is understood *not* to be the only materially feasible alternative but instead one selection from a vast range of possibility. While we routinely rely on automobiles, telephones, computers, ballpoint pens, and air conditioning, even our present limited scientific understandings would permit a wide range of alternatives to this ensemble, any number of which could comfortably assure our basic physiological needs. What is more, we can expect that each of these alternatives would offer its own unique attractions, as well as costs, when compared with present arrangements.

At a certain level this range of possibility is commonly recognized. Our contact with other cultures around the world, as well as with groups closer to home such as the Amish, clearly demonstrates that ours is not the only technological arrangement that can be made in the world. Stuck in yet another urban traffic jam, moreover, who among us has not yearned for the simpler life of our grandparents' time or for the connections with nature that we associate with Native American cultures before Columbus' fateful voyage. In the process, we at least implicitly recognize the possibility of alternatives. Yet we routinely dismiss such thoughts without genuine reflection. Whether by default or by design, the richness and actual extent of the range of technological possibility remains in our society unacknowledged, unappreciated, and unexplored.

There is, nevertheless, a necessary element of choice at least implicit in arriving at any particular set of technological arrangements. In the terminology of the contemporary study of science, technology, and society, it is said that technology is in some measure "socially constructed."[2] We might equally well label it "politically constructed."[3] The point in either case is to note that technology does not arrive of necessity or as the single-valued result of strictly material, "objective" conditions. Instead, it is shaped and, in effect chosen, under the influence not only of material, but of social and political conditions.

From this perspective, a commitment to democratic governance suggests that careful attention should be given to the precise mechanisms by which technological choices are made. In particular, it becomes important to be sure that the checks and balances traditionally in place to prevent undue concentrations of power in society are also effective over the full course of technology decision making. While responding to the will of the majority, we must also guarantee basic protections to minorities. If technology is not a neutral and apolitical phenomenon but something chosen, we must be sure that we are hearing from everyone and that particular, powerful perspectives on the world do not dominate the shaping of technology, setting the agenda and bounding political debate in such a way as to mute or exclude the voices of those who might dissent or seek to offer alternative perspectives.[4] Is it possible, in the extreme, that certain processes of technology development might afford the means for expression of certain values—i.e., of certain understandings of how we are to relate to one another and to the natural world—while failing to offer the rudiments of technological language to others, denying them expression at its most elemental level?

Also fundamental to the frame of thought adopted in this book, then, is the notion that truly democratic choice cannot occur without something approaching what political scientist Benjamin Barber has labeled "political talk."[5] We cannot make the claim of democracy without the kind of broadly based conversation among citizens that is, at its limits, unconstrained by preconceptions about what is or is not a "rational," sensible, or otherwise legitimate statement. In a democracy we must ultimately, as a matter of definition, show our willingness to hear each other out.

To put the matter more precisely, a true democracy attends to more than the easily heard voices of its most aggressive or most securely placed speakers. It listens actively for its more muted voices and for what may remain, even with them, as yet inchoate. Truly democratic conversation "places its agenda at the center rather than at the beginning of its politics" and carefully "scrutinizes what remains unspoken, looking into the crevices of silence for signs of an unarticulated problem, a speechless victim, or a mute protester."[6] True democracy is not satisfied with apparent acquiescence or superficial consent but actively pursues instead the actualization of equality of expression.

Conversations governing the shaping of modern technology have perhaps suffered more than others from domination by the loud noise of easily communicated material promise. This should come as no surprise: how can one argue with a program that promises liberation from "disease, hunger, and toil" and an enrichment of life through "learning, art, and athletics"[7] while sometimes imposing its most significant costs only in ways that are at best difficult to discern, at worst unknowable in advance.

At the same time, our lives may be more a reflection of the embrace of a particular selection of modern technologies than anything else. We have in a profound sense simply become, as Langdon Winner has so aptly expressed it, "the beings who work on assembly lines, who talk on telephones, who do our figuring on pocket calculators, who eat processed foods, who clean our homes with powerful chemicals."[8]

Where our more muted voices have gone unheard and this failure has had far-reaching effects on the lives of citizens through the design and configuration of modern technology, we must be concerned with the health of our democracy and with any ways in which how we treat each other might better measure up to the ideals articulated at the founding of this nation.

It is with concerns of precisely this nature that the two studies outlined in this book are presented. With a particular focus on the shaping of technology, this will prove above all to be a book about the importance of listening more carefully.

> "I will listen" means to the [true] democrat not that I will scan my adversary's position for weaknesses and potential trade-offs, nor even (as a minimalist might think) that I will tolerantly permit him to say whatever he chooses. It means, rather, "I will put myself in his place, I will try to understand."[9]

In a genuine desire for furthered understanding, this book flows from a struggle to hear the muted voices of those who would be participants in the shaping of their own technological futures but who have been silenced either actively or by their own reticence in the face of the eagerness of others to speak. Listening in this way, it seeks lessons that may be of general use in assuring the recovery of democracy in the shaping of technology for the future.

Two Studies

Too much of what we hear of opposition to nuclear power—or, in other periods, of opposition to technology ranging from textile machinery to genetic engineering—has been confined to narrow argumentation within essentially conventional conceptions of how we are to live in the world. Such argumentation has failed both to explain the commitment of opponents to particular new technologies and to communicate the concerns of those involved. It has failed even more fundamentally to call forth an exploration of possible expressions of alternative views through the realm of technological possibility and a formulation of technological possibilities that might serve as genuine alternative candidates in the construction of our collective future. Aspiring (as at least ostensibly we do) to live democratically, a greater effort has long been in order to gain a more complete appreciation for the views of our fellow citizens and for what the world may look like from the inside of movements urging alternatives to what others may regard as straightforward advances in technology.

In approaching the two studies to be presented in this book, however, the reader must be warned that the kind of listening contemplated here does not come easily or quickly. Where we

wish to come to an understanding of something that is beyond
routine experience, especially something that is not yet fully ex-
pressible for lack of developed technological as well as ordinary
linguistic resources, a few charts and graphs are not going to do
the trick.[10] Neither, of course, can my representations of the activities I will describe escape the effects of my own "selective attention,"[11] however vigorously I may try to report honestly and
without bias. The expression I give to the voices I have studied
will never ring true in its entirety, even among those whose perceptions and concerns I am working to convey.

With these warnings in place, I have begun both of the studies
presented in this book by trying to allow those involved to speak
for themselves. This has been particularly true in the case of the
first study, "the Alliance," in which I have begun simply with descriptions of the group's origins and functioning and with as rich
a description of the lives and backgrounds of seven individual
participants as I can imagine readers might tolerate. Wherever
possible, I have made extensive use of direct quotations. Actions,
however, are said to speak louder than words. And where Alliance
members have seemed short of the shared experience, technology, or ordinary words they might need to communicate their
perceptions and perspectives, a rich discussion of their backgrounds and activities has often seemed the best resource available in trying to achieve understanding.

The Alliance, like many grass-roots antinuclear organizations,
arose spontaneously from commitments deep enough to overwhelm the obvious difficulties one might expect in trying to sustain an organization without substantial funding or other
traditional resources. The depth of commitment of its members
is further evidenced by the success of its consensus governance
process, which has functioned with, if anything, greater efficiency
and effectiveness than the more traditional majority-rule or hierarchically structured organizations. In its time, the Alliance was
necessarily limited, to some degree, to reactive responses to authoritatively advanced technology. Both among individual members and as a group, however, it displays a surprising
sophistication in moving beyond classical models of political action, not only in its consensus process and extracorporate organization, but in its members' conscious avoidance of "top down,"
Washington, D.C., images of political action in favor of newly

emerging community-based approaches to political and technological change.

After presenting the Alliance as much as possible in its own
terms, I make an explicit effort at interpretation. Specifically, I
point to what I understand to be vigorous efforts to reconfigure
both interpersonal and human/nature relationships in ways that
come into fundamental conflict with the way these relationships
are structured through existing technological arrangements. Interpreted as part of a "value-oriented movement" in the context
of the study of collective behavior, Alliance activities leave us
with the somewhat startling image of conventional practice as a
form of extortion under which Alliance members are forced to
support patterns antithetical to their own deeply held sense of
how they should live in the world.[12] Our society's handling of Alliance concerns appears finally not only to violate basic principles
of democracy but to stretch the limits of even more rudimentary
notions of a social contract designed to guard against the simplest
forms of violence and abuse.

In the second study, it has been possible to describe home
power activities in a somewhat more conventional way from the
start because they have themselves achieved a greater degree of
coherence in terms of alternative technological and other more
conventional forms of expression. Interpretations in this case are
focused primarily on extracting from home power activities the
outlines of a general model for more democratic decision making
in technology design and development. This model begins from
the observation of "participatory research" and from a broad re-
claiming of agency on the part of the citizens involved in home
power efforts. What is observed is not greater citizen participation "in technology decision making" but greater citizen participation in general, as it happens to have touched upon technology
incidentally and on its own initiative rather than by official direction or arbitrarily bounded intent. The interest has not been in
technology per se but in living differently, with technology gaining attention only as a partial instrument.

A number of distinctive features of the home power movement
are described, including its emergence from a unique clearing un-
cluttered by the usually dominant institutions of our government
and economy. A new politics of participation closely echoes the
practices of the Alliance. And a partial withdrawal from the econ-

omy, somewhat resembling a reversal of historical transitions from subsistence to market economies, is described. The closeness of interactions among designers and users of home power equipment is also noted, along with explicit commitments to broad dissemination of knowledge and to "hands-on" participation by home owners, as evidenced in the field and in forums such as *Home Power Magazine.*

In formalizing home power as a model for more democratic shaping of technology, I end by drawing heavily on Benjamin Barber's notions of a "strong democracy"[13] in which "knowledge" is less the terrain of credentialed expertise than a basis for action that is arrived at democratically through "political talk." Opening conversation in unique and essential ways, Barber's strong democracy nicely describes in the broadest terms what home power has achieved in a specific instance, modeling the democratic shaping of technology. I end my theoretical interpretations of home power experience with speculations about the future of NIMBYism ("Not In My Backyard" resistance to technology) and by outlining some of the specific challenges we may confront in working toward a more democratic shaping of technology.

Both the Alliance and the home power studies, then, flow from an effort to listen more carefully and to give voice to views and perspectives that have not genuinely been afforded a hearing in our society. The first is presented with an emphasis on that voice and includes a minimum of explicit interpretation. The second is presented with more of a focus on trying to learn from home power practice how "hearing from people" might be built into the technology-shaping process more generally.

In reading both studies, but especially the ethnographic detail of the first, the reader may wish to keep an eye out for several recurring themes as a way of preserving links to the central concerns of this book. In both studies, there is evidence for a desire not for self-sufficiency but for a certain degree of *independence* denied under conventional technological regimes. In the first study, this is often a responsive desire stimulated by bad experiences with corporate and other seemingly arbitrary authorities. In the second, it is reflected in the development and adoption of actual alternative technologies. In both cases, it is associated with a desire to live differently or, in the words of one Alliance member, to "take part in building their own pictures, their own city and town, without feeling that big brother is coming in and doing it for them." In both studies, there is recurring evidence also of a desire for a more vigorous sense of community and for a reformulation of work roles that steps away somewhat from the traditions of specialization and corporate or contractual relations. And there is recurring evidence of action stemming from what would traditionally be called strong environmental commitments. Each of these themes is expressive of latent conflicts in value commitments that separate the individuals and groups I have studied from the technological mainstream. Taken together, of course, these departures accumulate into much more than alternative technology, although it is the expression of these departures from common experience and expectations through alternative energy technology that will remain the focus of my interpretations.

Again, it will remain for the reader to do much of the work to gather these themes, especially in the early stages of my presentation, as I will be trying to begin by allowing those involved to speak broadly for themselves.

Muted Voices

Part I: The Alliance: Alternative Value Commitments and the Shaping of Technology

1

The Alliance as a Group

THIS AND THE SUCCEEDING TWO CHAPTERS PRESENT THE RESULTS of an ethnographic study initially pursued in the early 1980s and focused on a (popularly termed) "antinuclear" group I will call "the Alliance." They will provide detailed descriptions of key participants and of the widely ranging antinuclear, prorenewable, antiwar, and other activities of the group's most active core of approximately forty people. The internal and external relations and consensus process of the Alliance will also be a focus for attention, drawing links with both the direct action and cash contributions (about ten thousand dollars a year) of several hundred individuals, virtually all of whom lived within the group's immediate rural community at the time of the study. As noted in the introduction, the initial objective in this account will be to relate honestly a story that participants themselves might have wished to tell; little will be done in this or the next chapter to draw explicit lessons or conclusions.

More specifically, this chapter opens with some further background on the study as a whole, outlining among other things, the state of energy futures debates in the early '80s. It then moves on to focus on group functions and the group's reliance on a consensus process. Chapter 2 will focus at an individual level, providing a detailed account of the backgrounds, attitudes, and beliefs of four active and three less active Alliance members. Chapter 3 will then provide a tentative interpretation of the study as a whole.

BACKGROUND AND CONTEXT

When this study was undertaken,[1] fundamental questions about the human implications of energy resource depletion had

23

received significant public attention in the wake of the disruptions of the OPEC (Organization of Petroleum Exporting Countries) oil embargo of 1973. The profound dependency revealed by oil and gas supply shortages and oil price increases around the time of the embargo and again after the 1979 Iranian revolution, as well as the social dislocations those shortages and price increases brought about directly, had generated arguments among responsible analysts that changes comparable to those of the Industrial Revolution were in store for modern societies.[2] Open challenges to "economic rationalization"[3] and to systems of economic growth (as contrasted with systems of sustainable production) were not uncommon outside official policy circles[4] and were often joined by implicit or explicit challenges to other fundamental elements of the "modernization trend" of the last eight hundred to one thousand years.[5] Linked with environmental sentiments of the time, and with actual or latent "decentralization," "appropriate technology," "voluntary simplicity,"[6] and other movements, the effects of the so-called energy crisis were seen by many to be of far-reaching consequence. One handling of these "broader issues," rare among official documents for its remarkable openness, was prepared by Willis Harman and others at Stanford Research Institute (SRI) International for the Energy Research and Development Administration* in 1977. Six major issue areas were discussed under the following headings:

1. National and global responsibilities

2. Trade-offs determining the socially desirable level of economic activity

3. Needs for and costs of government regulation

4. Centralization vs. decentralization

5. Predominant societal goals

6. Rationality governing decision making[7]

In addition, three perceptions (labeled "A," "B," and "C") of the world and of the debate over energy futures were outlined, each of which was thought to have "significant representation among

---

*The ERDA was the short-lived precursor to today's Department of Energy.

decision makers, analysts, and the body politic"[8] at the time. These perceptions span a broad range. At one extreme (Perception A), firm support for a "high-growth, high technology, free enterprise society" is accompanied by a belief in the efficiency and effectiveness of centralized management within a framework of economic rationality. Resource limitations are not seen to impose limits on past patterns of technological innovation, and the future, with respect to energy and other matters, is seen largely as an extrapolation from the past. At the other extreme (Perception C), finite resources and a sense of fairness to other nations and to future generations require that developed societies move rapidly away from the basic paradigms of industrialization and toward a new "planetary stewardship." Under this alternative view, the attractiveness of fulfilling work and other nonmaterial goals far outweighs that of extended growth, while a move toward decentralization and a break with the dominance of economic rationality in decision making is thought to be very much in order.

Using the six issue areas listed above, SRI's three perceptions are summarized in somewhat greater detail in the accompanying table. Further review of these three perceptions, even considering only one of the six issue areas, reveals very different understandings of the role of energy in society, and very different ideas as to proper socio-technical solutions to society's energy problems. Yet Harman and his colleagues at SRI conclude that all three views

> have to be honored in some sense since each "fits" the observations of his environment as made by the person holding that view. Presently there appears to be no clear way of standing apart and objectively determining that one of the views is more "true" than the other two. No one of these three views can be disproven by "facts" drawn from another.[10]

Moreover, according to Harman and his colleagues, "Whichever view society comes to accept, that view will tend to become more 'real.' "[11]

Prominent as they were, energy issues were by no means alone in distinguishing the period of the middle 1980s. In understanding the activities of the group to be described here, it will be important to note that this was also a period in which nuclear armed

Characteristics of Three Perceptions[9]

WORLD RESPONSIBILITY

Characteristics of Perception A. The benefits of a high-growth, high technology, free enterprise society as compared with any feasible alternative are obvious and generally agreed to. Such a society provides the best hope for raising the nation's poor, and the poor of the world, to a higher state of material and social well-being. Hence a U.S. responsibility with regard to the world is to maintain its technological and economic leadership, and to aid poorer nations to industrialize and modernize.

Characteristics of Perception B. A "fairness revolution" is necessary; the rich nations consume far more than their share of the Earth's limited resources and contribute far more than their share of environmental damage. The insistence of the poorer nations on a "new international economic order" is justified; a new order is essential to any hope of eventual world political stability. Richer nations like the U.S. need to consume less, support the redistribution of resources, and recognize the validity of societal choices other than Western style industrialization and agribusiness.

Characteristics of Perception C. A "fairness revolution" in the world is necessary; however, this cannot come about without fundamental change in the nature of modern society. The basic paradigm of industrial society contains the seeds of international confrontation over finite planetary resources. It contains no rationale or incentive for planetary stewardship or for more equitable distribution of the Earth's resources. The industrialization trend and the goal of material progress, in the absence of more transcendental values, lead ineluctably to problems of resource depletion, environmental deterioration, hazardous substances, threats to the planet's life-support system, and international competition for the means to survive. Thus the desirable goals of Perception B cannot be achieved without a fundamental transformation of industrial society and its institutions.

ECONOMIC GROWTH

Characteristics of Perception A. Economic, material, and technological growth must continue at least for the present. There will eventually be a gradual leveling off of some aspects of economic growth and resource usage. But those neo-Malthusians who see crippling shortages in the near future are ignoring our demonstrated ability to cope with such problems by technological innovation, substitution, exploiting lower grade ores, etc. When these factors are taken into account, we are nowhere near ultimate resource limits. The social costs of abruptly lowering economic growth rates, particularly in terms of unemployment and dooming the poor to hopelessness, are too high to consider.

Characteristics of Perception B.   The social and environmental costs of continued material and technological growth in the pattern of the past, and of continually rising per capita energy consumption, are intolerably high. There must be a leveling off. There is a "new scarcity"—of physical resources, waste-absorbing capacity of the environment, resilience of planetary life-support systems—qualitatively different from the scarcity problems "solved" by modern industrial production. The response to this must be a voluntary choice of simplicity and frugality in the lifestyle of the society, and a renunciation of the consumption ethic and the growth addiction.

Characteristics of Perception C.   Fulfilling work is necessary to the well-being of persons. The trends of increasing industrialization and automation, however, lead to less and less intrinsic reward in work. Environmental and resource depletion impacts put constraints on expanding production and hence on employment. Thus, there develops chronic unemployment and underemployment. These worsening trade-offs are not accidental, but intrinsic to advanced stages of a society that evolves around the present industrial paradigm.

REGULATION

Characteristics of Perception A.   Regulation will be increasingly necessary. That is part of the price we pay for the benefits of highly complex modern society. But regulation should be applied with moderation; we must not kill the goose that lays the golden eggs. A central concern of the federal government should be to do the necessary regulating in such a way that business has the stable climate it needs to maintain a healthy economy and risk new technological ventures.

Characteristics of Perception B.   The only way to avoid regulation that will ineluctably infringe on civil rights and liberties is through reduced demands on resources, and simplification and decentralization of society.

Characteristics of Perception C.   The regulation dilemma, too, is perceived as a consequence of the playing out of the industrial paradigm. Consequently, the attempt to continue past growth patterns and then control the negative consequences is a losing game. In general, the higher the level of energy use (or economic production) the greater the need for regulation—and hence the greater the threat to civil rights and liberties.

DECENTRALIZATION

Characteristics of Perception A.   On the whole, centralization promotes efficiency and effective management. We need to alleviate any problems that excessive centralization appears to cause, and in particular we need to limit centralization that tends to eliminate competition. But it is important to be realistic. Some of the decentralization talk is romantic nonsense; we can never go back to the family farm.

Characteristics of Perception B.   Decentralization of control, of technology, of population, is necessary if we are to have a humane, just, and free society in the future. In particular, this implies "appropriate technology" and the strengthening of community.

Characteristics of Perception C.   Decentralization is seen as one of the essential characteristics of the post-transformation paradigm. Decentralization of technology, population, production, control and mass culture can foster the values of self-determination, community, ecological stewardship, equity, and liberty.

GOALS

Characteristics of Perception A.   The ability to achieve humane goals has been, and will continue to be, made possible through economic growth and material and technological progress.

Characteristics of Perception B.   Humane ends can be attained only by putting more emphasis on fostering human growth and development and reducing the emphasis on material accomplishment, on consumption, on economic growth as an end in itself. The goal is to be, rather than to have or to control.

Characteristics of Perception C.   Like Perception B, this view sees a resurgence of emphasis on humane goals. Industrialized society is increasingly unable to provide goals that will enlist the deepest loyalties and commitments of its citizens. Inherent in the post-transformation paradigm is a reassertion that man has two complementary aspects, the physical and the spiritual, neither explainable in terms of the other nor to be denied because of evidence for the other. Within the spiritual aspect, which the rising prestige of positivistic science for a time tended to debunk and confute, are to be found the proper goals for the individual and society. Thus, the transformation that is perceived to be in process goes (as in Kuhn's use [1970] of the term paradigm) to the level of the basic vision of reality around which society's institutions are constructed.

Decision Making

Characteristics of Perception A.   Decision making in society needs to be made on as rational a basis as possible. There can be little doubt that quantification of data and the development of the conceptual frameworks of economics have contributed greatly to this desired objective.

Characteristics of Perception B.   Decision making in society needs to be made as participative as possible, and to be guided more by humane criteria. The new mood is a reaction against the predominance of materialistic values; against tendencies to put human affairs and even lofty goals in economic terms, to assume that the wisest decisions are made on an economic basis, and to elevate means—production and consumption of goods and services—to the position of ends. It is necessary to judge a social choice, not in the narrow sense of profit and efficiency, of growth and productivity, but in the broader sense of social efficiency—does it pollute the environment, squander resources, bring about disemployment, stultify workers, misguide consumers?

Characteristics of Perception C.   The domination of decision making by economic rationality is an intrinsic aspect of the industrial paradigm. This characteristic, and the problems it engenders, will change only when the paradigm changes.

cruise missiles were being deployed in Europe by the Reagan administration and there had as yet been no hint of the impending fall of the Berlin Wall or the almost equally abrupt end of the cold war. This was a time in which the Three Mile Island and Chernobyl nuclear reactor disasters remained very much in people's minds, and new nuclear power plants (e.g., the Diablo Canyon reactors at San Luis Obispo, California) were still being brought online by major electric utilities. President Carter's pre-1980 efforts on behalf of conservation and renewable energy development had in many cases been precipitously curtailed by the Reagan administration. And the environmental policies of Interior Secretary James Watt seemed to be on the ascendancy with "environmentalists" widely vanquished or struggling to hold their own.

In this context of potentially historic tensions and uncertainties, the interests and commitments that inspired this study were similar to those Robert Coles describes at the beginning of his monumental three-volume series, *Children of Crisis*:

> I am working under the assumption that there still is room—maybe a corner here and there—for direct, sustained observation of individual human beings living in a significant and critical period of history. By direct observation I mean talking with people, listening to them, watching them—and being watched by them. By sustained observation I mean taking a long time: enough time to be confused, then absolutely certain and confident, then not so sure but a little more aware of why one or another conclusion seems the best that can be argued, or at least better than any other available.[12]

As a complement to the ordinarily remote policy handling of groups such as the one to be described here, an emic or "insider's" view was sought and an ethnographic approach not unlike that suggested by Coles was adopted as the most appropriate way to develop such a view.

The basic hypothesis was, in effect, that a close ethnographic look would reveal unjustifiable simplifications in official policy treatments of energy just as it has, for example, in the area of narcotics addiction:

> Policy—and the treatment, prevention, education, and law enforcement efforts that it logically implies—is built on assumptions of what "those people" [i.e., narcotics addicts] are like. Yet a little ethnogra-

phy quickly teaches you that the assumptions, at their very best, are oversimplifications.[13]

The object was to ascertain in at least one specific and concrete case, what "those people" actually were like.

Motivated in this way, and with our frame of reference now established, this chapter's examination of shared or collective aspects of group activities is divided for expositional purposes into consideration of the Alliance's setting and origins, its objectives, its activities in support of those objectives, its consensus approach to decision making, and its relation to the community in which it has operated.

SETTING AND ORIGINS

The group I refer to here simply as "the Alliance" came into being in 1978 in general opposition to the development of nuclear power and in specific opposition to the opening of the Diablo Canyon nuclear power plant in San Luis Obispo, California. During the period of these observations (1981–84), the Alliance drew its participants primarily from the area of three small towns at the outermost fringes of the commuting range of a major coastal city. The area is sparsely populated and at times strikingly beautiful in its combinations of Pacific coastline, large natural areas (including several major parks), and extensive dairy operations with their vast expanse of large and nearly treeless hills turning from a lush springtime green to a California gold in the summer. Statistically, the community[14] is rather affluent. Its dairy operations are prosperous, though small-scale in the sense that each is basically run by a single family, and a good portion of its remaining population is seasonal, consisting of city people with second homes in the area. The towns themselves consist of little more than a post office, small grocery store, and an expensive shop or restaurant or two primarily serving the flow of fair-weather (summer) tourists.

Although loosely affiliated through the "Abalone Alliance" with approximately sixty other groups throughout California, the Alliance and its activities were initiated by members of the community and not through any organizing effort on the part of the Abalone Alliance.

As an important facilitator of the group's first meetings, one of the group's members, known here as Jane,* offers insights of both a personal and a collective nature into the group's formation. Jane had been trained as a community organizer in the Chicago area as a part of her work with rape victims there some time after she finished college. Her training had been at a school that "didn't want to train professional organizers," but instead "wanted to empower people who were functioning as organizers and who did not have enough skill to do that effectively." Eventually employed in the Alliance community as a waitress, Jane drew no income from her work with the Alliance at the time of the group's formation or thereafter. (Her primary role in the group came to be running occasional all day nonviolence training sessions for the Alliance, training members of the community who intended to participate in nonviolent demonstrations and/or civil disobedience.) As an individual, Jane's participation in forming the Alliance was stimulated in part by finding in a major newspaper a picture of a friend who had been involved in an important antinuclear demonstration on the East Coast.

Even before the first meetings of what was to become the Alliance in the early fall of 1978, discussions among residents of the community had revealed a shared interest in doing something locally. The first major demonstrations at the Diablo Canyon nuclear power plant had occurred around this time, and one of the Alliance's earliest major community events, which took place in October 1978, included audience participation in conversations with thirteen protesters from a neighboring town who had recently been arrested at Diablo Canyon. "So it was up in the consciousness of everybody that something was going on. Something was building." As Jane put it, "We met; we decided we wanted to do it, that there was something to do; we found ourselves a name;" and the Alliance had, in effect, been formed.

OBJECTIVES

Unconstrained by the communications and control difficulties of larger and more formal organizations, the Alliance has been

---

*Pseudonyms preserving gender and referring to actual individuals will be used in place of real names throughout this study report.

free to pursue an ever-evolving collection of objectives. Any discussion of Alliance objectives must in this sense begin with recognition of the close working relationships of active members. With few exceptions, those forty or so active members who attended the group's meetings in the early '80s had known and worked closely with each other over a period of years. The group did not live communally (nor did any subset of the group) and in fact rarely managed to get together for purely social purposes; yet its members shared much hard work against what they saw to be heavy odds and had in this process established what amounted to their own small and somewhat isolated alternative society. Mutual understanding and trust[15] ran high.

For this and other reasons, there was generally little need to state the objectives of the group in explicit terms. Objectives were implicit in the work, understood by members, and free to evolve with little danger of confusion or misunderstanding. Quite conscious of this aspect of their activities, the group, in fact, actively chose on at least one observed occasion *not* to define its objectives more explicitly. One member had, on this particular occasion, asked that the group agree to a stronger affirmation of its interest in renewable energy systems. The group was uncomfortable not with the renewable energy objective but with the explicit definition of objectives, which might, some felt, become a tempting "lever" for pressuring people into particular activities. There was a preference expressed that goals and objectives continue to be defined to a large degree implicitly through the activities and events that members felt a commitment to support. Furthermore, the activities of the Alliance were so thoroughly rooted in the community as a whole that a description of the Alliance or statement of its objectives was rarely needed even for external fund-raising purposes.

At least two descriptions of the Alliance and its objectives nevertheless remained in the group's files as of 1983 (files that were maintained by the group's treasurer). The first, prepared a full year after the group got started, offers a number of details of the group's early activities as well as an indication of the development of its objectives.

> We're a group of [local] citizens who joined together around our common concern about the dangers of nuclear energy. During the year we've existed, we have also come to focus on the need for alternate energy sources, such as solar power.

Other nuclear-related concerns are mentioned—e.g., plutonium leakage from nuclear materials dumped in the ocean, the Rancho Seco power plant (also in California, and "a twin of Three Mile Island"), and the Lawrence Livermore Weapons Lab, east of San Francisco. Noting that, "We'll be part of a nonviolent blockade of Diablo, and are ready to go to jail if necessary to express our concern about this threat to all life," the description also refers to the Alliance as a "safe energy group . . . recently . . . alarmed by the prospect of oil drilling along [the] coast." Reference is made to the group's support for a proposal to convert a closed military airfield into a "solar village" and to a solar water heater built and installed by members and friends at a local community center. In its first year, the Alliance is reported to have "sponsored a half dozen public events . . . in the community, . . . distributed literature, spoken to [a number of] groups, [placed] articles and letters in the paper, and . . . opened an office in town." (This office was maintained continuously from its opening at least until the fall of 1987.)

In its somewhat more elaborate three-page "Review of Activities" dated May 1982, the Alliance listed its purpose as follows: "[T]o educate the local community about nuclear energy and weapons, to work against the construction and operation of nuclear power plants, and to promote safe, renewable sources of energy, including conservation." The group's activities were then reported under six headings: "Opposing Nuclear Power," "Promoting Renewable Alternatives," "Preserving the Coast," "Preventing Nuclear War" (an area of activity that developed after the 1980 election of Ronald Reagan), "Newsletter," and "Regular Meetings."

ACTIVITIES

Among Alliance activities, "Regular Meetings" figured prominently throughout the period of this study. These occurred every two to three weeks almost without exception and involved at least five consistent participants on average, with exceptional attendance as high as thirty or so. It is through these meetings, generally lasting about three hours and held in the evening at various members' homes on a rotating basis, that the Alliance selected and planned its other activities and made or delegated its decisions through a sophisticated consensus process. (More will be said about this consensus process in a later section.)

Other Alliance activities might somewhat artificially be considered in three groups: (1) activities that informed and/or directly involved the community, (2) activities aimed at maintaining an informational network with other individuals and organizations both inside and outside the local community, and (3) activities involving "direct action."

The first of these groups prominently included maintenance of the office itself. A one-room structure centrally located on the main street of one of the area's three small towns, it was rented on a monthly basis by the Alliance and kept filled with books, newsletters, magazines, and other informational materials covering all aspects of the group's areas of interest. The office also offered posters, buttons, bumper stickers, T-shirts, and other items of this sort for sale. Major community-centered activities in addition to the office included the Alliance-organized solar fair of 1980, which included (to give an idea of its scale) three photovoltaic manufacturer displays, a full-size wind generator, and attendance by such organizations as Coevolution Quarterly. In another major community event in 1984, the Alliance joined forces with the local community chorus to produce a musical play called *Blue Line*,[16] which had been written for the purpose by a regular Alliance participant and involved the efforts of approximately fifty people. Performed six times (three within the community, three in a neighboring community) for audiences of about forty people each time, the play followed the amusing but informative exploits of an affinity group* from the fictitious town of Esperanza as its members participate in the 1981 blockade and civil disobedience at Diablo Canyon. Perhaps ten times a year, the Alliance organized and hosted smaller events at a community center seating approximately 100; these events included films (e.g., *The Last Epidemic*, a film about the impacts of nuclear war), speakers (including both local people and, occasionally, well-known figures such as Ted Taylor**), and other activities (e.g., a

---

*"Affinity group" was the term commonly used to describe a group of protesters who had received nonviolence training and traveled as a group to participate in nonviolent protest actions together.

**Taylor was the subject of John McPhee's book, *The Curve of Blinding Energy*.

"Peace Charette"[17] in which local citizens gathered to brainstorm and discuss local opportunities to work against nuclear war). All of these events, which together constituted the largest fraction of the group's efforts, informed and involved members of the community.

The second cluster of activity involved maintaining contact with a network of other individuals and organizations, both within and outside the community. The closest ties were, as might be expected, with other groups centered in the community, including planning, conservation, bike path, and other organizations. At the conclusion of this study, the strongest of these ties, perhaps, were with a local section of the Women's Party for Survival, which worked almost exclusively on peace and disarmament issues and actually drew many of the women out of Alliance meetings who had been active in the Alliance before the Women's Party was formed; the Alliance and the Women's Party frequently cosponsored activities and were always in close contact with each other. Strong ties were also maintained with a large number of groups outside the community, often through personal acquaintance with representatives of those groups. These acquaintances were in some cases established through joint actions far removed geographically from the Alliance's home base (e.g., through Diablo Canyon demonstrations) or through correspondence on specific issues that particularly affected a distant group. Ties of this general nature had been established with a group opposing work on nuclear submarines in Washington state, with the initiator and participants in a transcontinental "Walk for the Earth," with a national networking organization trying to stimulate response to the dangers of nuclear holocaust, environmental deterioration, and human oppression,[18] and with many other groups and organizations. In addition, the Alliance maintained memberships in and received a continual stream of communications from a host of nonprofit organizations: the War Resistors League, Environmental Action, the Fellowship of Reconciliation, and Common Cause, to name only a few. Fifteen minutes or so of each meeting were typically devoted to the distribution of the stacks of magazines, newsletters, appeals, and other mail accumulated between meetings to interested/responsible members.

A final but very important element of "networking" activities at the Alliance was its relationship with its state and federal representatives, executive agencies, and other established interests. The group held frequent letter-writing "parties," and several members generated letters to the editor on a regular basis. A smoothly functioning phone tree was used to alert the community as a whole when a strong response was needed quickly on a particular issue. Its emphasis on local action notwithstanding, the Alliance was by no means entirely alienated from the conventional political process; individual members frequently talked over the phone with or met directly with their representatives or other officials, and the Alliance often received unsolicited and highly personalized mailings, sometimes from sympathetic officials seeking nonmonetary support for specific legislative efforts. With the approach of the 1984 national elections, there was clearly a strong interest among individual members in seeing Ronald Reagan defeated, preferably by Senator Gary Hart, although the Alliance itself never appeared to become strongly or directly involved in political campaigns for office. (The Alliance did, on the other hand, directly support efforts to place the "Nuclear Freeze," "Nuclear Free California," and other initiatives on the state ballot.)

The third and final set of Alliance activities has been loosely termed "direct action." Included here, of course, are the marches and demonstrations generally associated with this term. Depending upon how one defines membership in the Alliance, by 1984 somewhere between one-quarter and half of the members had been arrested at least once at direct actions of this sort, and probably more than three out of four had participated in a manner that involved their physical presence. Demonstrations and marches of this kind ranged from an Alliance-organized Hiroshima Day "vigil" in which about twenty-five members lined up in front of the Alliance office holding a banner reminding residents and weekend tourists of the occasion, to participation with thousands of other people in actions for peace (e.g., in the city) or against the opening of Diablo Canyon in San Luis Obispo. On average, the Alliance was represented by some group of its members at between ten and twenty major out-of-town demonstrations each year.

Other instances of "direct action" included the construction of a solar water heater for a community center mentioned earlier and the addition of a solar greenhouse to the Alliance office, accomplished by members in 1981. As of 1984, money had also been set aside for the purchase of a photovoltaic electric system to power lights for the office. (The office always had phone and water connections, but it was heated with a small wood stove and had never been connected to utility lines for electric power.) The Alliance also did some research on its own; one active member, for example, carefully explored the full range of nuclear activities in the state of California (weapons, energy production, materials handling, materials storage, and materials dumping), depicting his results in a colorful and informative poster that was printed and sold at various events and in the office. In another form of direct action, the Alliance often supported members and friends in individual efforts that furthered Alliance goals. Such support was, at various times, provided through direct dollar grants (as occurred in connection with the *Blue Line* play, for example), loans (as occurred in the local professional production of children's antiwar songs that were played free on major area radio stations), and other forms of assistance. (As one further illustration of this sort of support, a member of the community who did not otherwise participate actively in Alliance activities once used the Alliance mailing list to raise $892 to help pay for his travel to Nicaragua under the auspices of Oxfam America.)

As has already been noted, the categories employed here in describing Alliance activities—i.e., meetings, community information/participation, networking, and direct action—are somewhat artificial. Many of the activities described would fit with equal ease in more than one of these groups, and even where they suggest conscious and specific Alliance intentions are implied (as might often be the case, for example, under "community information/participation"), they do not adequately reflect the integration of objectives generally observed in practical situations. With respect to intent, only one distinction was raised specifically with any frequency: the distinction between events/activities intended primarily to *serve* the community and the objectives of the Alliance and events/activities designed primarily as *fund-raising* functions to make other efforts possible. The Alliance typically raised and spent between five thousand and ten thousand dollars each year exclusive of salaries. (The Alliance periodically maintained one half-time staff person—more on this in a later section on "Participation and Relation to Community.") Fund-raising was therefore a consideration in many or most Alliance activities.

At virtually every community center function, for example, there were tables at the entrance asking for contributions and offering T-shirts and other items for sale from the office stock. Donated food was also sold at other tables during breaks in the program. Several times a year major efforts specifically aimed at fund-raising were also mounted. A number of auctions (items donated), benefit concerts (sometimes in neighboring communities), or other events raised more than one thousand dollars in a single day. Whenever an event afforded an opportunity for unusually successful fund-raising, its planning during an Alliance meeting tended to include an explicit clarification of "service" and "fund-raising" priorities in the process of deciding, for example, between a grant and a loan for a particular project, or on the "requested donation" for community center events (between zero and six dollars over the course of these observations).

Consensus and Mutual Support

The Alliance invariably acted not by majority vote but by consensus in a process that has long been employed by similar groups throughout the country.[19] Decisions were made, or in some cases delegated to individuals or subgroups, during regularly scheduled meetings twice a month; no proposal became effective, however, (i.e., no decision was made) until *all* those present at the meeting agreed to it. Although well known in some circles, this consensus process and its influence in the course of Alliance meetings is probably worthy of at least a brief description here.

The consensus approach, both in general and as practiced by the Alliance, places an emphasis on hearing from everyone. This was clearly evidenced in Alliance discussions by the periods of silence that were allowed to occur and by the explorations of eye contact and direct questioning aimed at those who had not spoken. It was also structurally supported through the use of a facilitator—a role that was taken in rotation from one meeting to the next by virtually all participants—rather than a fixed chairman or leader. As particular issues were resolved in discussion and decisions began to emerge, the facilitator would ask, "Are there any reservations?" in a further attempt to elicit differences that might otherwise have gone undetected. Only when this question and a noticeable pause produced no further discussion—i.e., only when everyone had contributed views and ideas and expressed no continuing opposition to a proposed course of action—was there understood to be a consensus and a decision on which the group could act. With few if any exceptions, individual commitments to this priority of true consensus over decision and action were striking, both in the handling of specific issues and in the sophisticated and smoothly running procedures that had been developed for meetings.

Alliance meetings began with the setting of an agenda. Topics for the group's attention could be suggested by anyone present and were usually listed by the facilitator on a large sheet of paper for all to see. Estimates of how much time would be needed to cover each topic were also noted. These topics and time commitments were then adjusted as necessary until the agenda was acceptable to all present, the only recurring adjustment being to postpone one or two topics to a future meeting when total time commitments appeared to exceed about three hours.[20] Including both routine and special, informational, decision-oriented matters, a typical agenda (such as the one below) took only about five or ten minutes to develop.

minutes	*agenda item*
10	Peace Camp
5	Diablo Update
7	Greenham Women
5	Nuclear Free California Initiative
5	Walk for the Earth
15	Mailbag
12	Bank Account
10	Break
10	Announcements
2	Treasurer's Report
5	Blue Line Play Update
10	Blue Line Dollars
2	County Peace Center Appeal
15	Closing

(Note that time estimates were initially conservative—this agenda actually took about three hours to get through.)

As meetings proceeded, a timekeeper, usually someone other than the facilitator, let the group know when time allotments for specific topics had expired and the group rapidly (generally within a minute or two) decided, again by consensus, how much if any more time they wished to "contract" for that topic. Informational/individual topics were pursued until all interests/questions had been answered or satisfactorily deferred. From the listed agenda, for example, one member described a distant peace encampment at a weapons lab and his plans to travel there, inviting any who might be interested to go along with him. In another example from the listed agenda, information from a neighboring community regarding a local visit from one of the "Greenham Women" (a group of English women resisting President Reagan's cruise missile deployments in Europe) led to a volunteer who agreed to explore specific dates and locations for a possible Alliance-sponsored event.

When decisions were required, the meeting went forward only after the question, "Are there any reservations?" brought no further response. While it was not unusual to hear phrases such as "I think I would have to block consensus if . . ." during discussion, and while a decision was sometimes postponed to another meeting, consensus was almost never blocked, and business appeared to be conducted at least as expeditiously as it might be under more conventional parliamentary rules.

Every Alliance meeting ended with a "closing." With everyone seated in a circle insofar as possible throughout a meeting, the closing consisted of going around the circle to hear any closing remarks each individual might wish to make. While people rarely took more than a couple of minutes, there was no time limit explicitly set. Neither were closing remarks confined as to subject (topics might be entirely unrelated to the substance of the meeting) except that they were never supposed to respond to or answer, in any direct way, anyone else's closing remarks. As specific issues were left behind, the closing often seemed to reaffirm the unity of the group, frequently with greater humor or emotion than had been evident in the business portions of the meeting, while at the same time carefully providing each individual with an unchallenged opportunity for self-expression.

The consensus process employed by the Alliance was only one manifestation of a broad tendency toward mutual support in the group. This support was evident on a large scale as the whole group rallied behind major projects such as the *Blue Line* play or the solar fair and the individuals who had initiated them. While particular interests and approaches vary among individuals as these two examples initiated by different individuals indicate, these differences appeared to be heavily outweighed by the shared commitments of all members; in practice, major projects almost seemed to involve a planned convergence of energies around the central interests of individual members, each in his/her turn. On a smaller scale, the group was highly sensitive to the depression and "burnout" individuals often experienced as their efforts appeared to vanish without a trace against the enormity of the forces generally arrayed in opposition to the group's efforts. Support and encouragement—indeed appreciation—was clearly expressed by the group for the participation of each member. (A remarkable aspect of this appreciation was that it never seemed to be used as a lever. The group welcomed all participation in its efforts but did not appear to demand or even expect any specifiable level of contribution from individual members. Members contributed what they were comfortable contributing, and a very wide range in the level of effort expended by individual members on behalf of group objectives was both acceptable and expected.)

In response to questions about the origins of the consensus process in Alliance activities, one of the group's founders (referred to earlier by the name Jane) indicated that "it was part and parcel of what those founding members brought with them into the organization." While members of the Alliance were able to learn much with respect to specific techniques from other sources over the years, most appear also to have begun as individuals with an unusual inclination to listen and to appreciate the efforts and commitments of others. (More will be said about individuals in the next chapter.)

Participation and Relation to Community

Close ties between the Alliance and the surrounding community were maintained as a matter of some importance. This should already be apparent to some degree through earlier discussion of community events, ties with other community organizations, and other aspects of the Alliance and its activities. In this section, specific attention is devoted to the ways in which members of the community actually participated in Alliance activities, to the particular role of a few active members in maintaining contact with the community at large, and to the particular exchanges involved when the community periodically supported a half-time Alliance staff person. As each of these topics is presented, the reader may wish to consider how they may represent extensions of the group's internal consensus and mutual support processes.

Participation or "membership" in the Alliance was not accurately reflected by attendance at regular meetings. The full mailing list (maintained on one active member's home computer) in fact contained about nine hundred names and addresses, and unusual occasions sometimes brought a positive response from as many as four hundred or five hundred of these. As an example, shortly before the Diablo Canyon power plant was allowed to begin low-power testing in 1984, approximately four hundred of these "members" contributed a total of more than two thousand dollars through an Alliance telephone campaign to place informative ads in local papers and mount a final effort to block the plant's operation. Many "members" in the community were relatively invisible because they only attended Alliance-sponsored community functions or made special contributions to specific projects; one member the author took lodging from for a couple of months, for example, was not known to have attended a single Alliance meeting but gave about one hundred dollars to the *Blue Line* production and later attended the play. More broadly, large numbers of people in the community who did not attend meetings at all functioned routinely in support of the Alliance. "Jane," as mentioned earlier, conducted nonviolence training sessions. "Bob" started seedlings and generally watered plants and for several years maintained the greenhouse at the Alliance office. Several semiprofessional singer-songwriters were essentially available on call for benefit concerts and other functions. Other members specialized in preparation of artistic banners, signs, posters, etc., or in running the food concession at community events or in other capacities, regularly filling Alliance needs. Much of the group's efficacy and continued smooth functioning, in fact, was clearly due to the intimacy of its relations with the community and to a closely cooperative integration of skills and interests in a larger circle of participants. This integration was accomplished, it should be emphasized, through a continuous exchange between and among those who did and did not attend meetings regularly: those who did not attend meetings were by no means simply assigned tasks by those who did.

Several of the most active members of the Alliance were also uniquely situated in the community in a way that greatly facilitated the kind of communication necessary for the group to function as it did. One, referred to in later sections as "Brad," ran his own window-washing business until roughly 1981 and then started his own appliance repair service. Another was the only utility-certified energy conservation contractor in the area. Both of these people interacted frequently with a fairly broad cross section of the community and came to know a large proportion of the community (whose total population, again, was quite small) personally over the years. As of 1983, an Alliance staff person also actively cultivated contacts throughout the community by manning the office, in his publicity efforts at the post office and local papers, through actual Alliance events, and in other official and personal capacities. With a background in social work, this person also worked for the county in all the area's school systems. The previous staff person, who moved out of state, was similarly involved with the community and also worked in a local bookstore. These active members served a special function, then, both inside and outside regular meetings by providing coordination and communication channels for the larger participant group. In several cases, it should also be noted, they were in a position to interact not only with sympathetic and potentially sympathetic members of the community but also with people who clearly did not support the group in any of its efforts.

Monetary support for the Alliance was derived almost exclusively from the community itself. Rather than being supported by sources outside the community, it was itself a source of support for outside projects and organizations, sending hundreds of dollars a year, for example, to its statewide parent, the Abalone Alliance. This also was the pattern in a completely separate fundraising and bookkeeping effort when the Alliance had a paid staff person. Averaging one half-time position and ranging between no position and one full-time position, paid staffing occurred when a member of the group obtained the group's permission to try to and then did raise his/her own salary in the community. This was first accomplished by "Frank," beginning in the fall of 1981. Wishing to devote more of his time to Alliance activities than his job at a local bookstore allowed, Frank began by simply writing letters to members and friends in the community describing the projects he wished to pursue and asking for support. He initially raised only about two hundred dollars per month to support himself part-time as a "coordinator" for the Alliance. Then in late 1981, with a desire to work more intensively on peace and disarmament issues, he sought support for a full-time commitment. Noting in his first letter, "My vision is that every community of concern . . . would support one of its members to work full-time on this task of global survival," Frank successfully raised about fourteen thousand dollars from seventy-seven contributors in 1982 and a similar sum in 1983. (A bit more than half of this amount he took as salary, the remainder as expenses including office rental and supplies, travel, etc.) He kept contributors and others informed of his activities through written quarterly reports. After Frank's marriage and departure from California, "Bill" followed a similar procedure in raising funds for his efforts writing and putting on the *Blue Line* musical play,[21] staffing the office, and serving as a coordinator for Alliance activities. When one considers the size of the community and the number and scale of contributions associated with this approach to paid staffing, it should be clear that the interactions involved were an important part of the life of the community and a major factor in the closely participatory relationship between the Alliance and the larger community.

SUMMARY

As an informal organization, the Alliance was formed, sustained, and operated on the initiative of local citizens. From 1981 to 1984, its actions locally and at the state and federal levels relied on close communication among its members and were based on strongly shared opposition to nuclear power and off-shore oil development, as well as support for conservation and renewable energy alternatives and a concern regarding nuclear weapons and the potential for nuclear war. The group was not hierarchically structured and showed a strong practical commitment to a consensus process in its decision making. Direct, personal action was consistently the most common expression of the group's concerns. Although this chapter has described only the collective functioning of the Alliance, alternative value commitments with respect to person-to-person and man/nature relationships should already be apparent in outlines that will remain consistent with the group's alternative technological preferences.

2

Alliance Members as Individuals

CHAPTER 1 HAS BEEN DEVOTED TO A DISCUSSION OF THE ALLIANCE IN its collective manifestations. A complete understanding of the group and its activities is impossible, however, without a more detailed consideration of at least some of the group's members as individuals.

After a brief background discussion, therefore, this chapter is entirely devoted to a description of four active and three less active Alliance members. Chapter 3 will then offer one possible set of interpretive conclusions.

BACKGROUND

From 1981 to 1984, the population of the community within which the Alliance worked could loosely be divided into four groups: (1) wealthy summer residents, (2) modest- and low-income year-round residents who prize the natural and social/intellectual setting, (3) ranchers, and (4) drifters. The drifters were a highly visible though small fraction of the population during the summer and were perhaps attracted by the mild climate, the relatively relaxed attitude toward sleeping by the road, the beach scene, and a certain trickle down of money and seasonal work from the wealthier residents. The first and second groups above were the most closely tied socially, with the ranchers fewer in number, quieter, and to some degree isolated from the other three groups in their interests and social pursuits. The preponderance of those who were active in the Alliance (those who did more than occasionally attend an Alliance-sponsored community event) could best be characterized as members of the modest- and low-income year-round resident group.

Active members of the Alliance were typically in their thirties or forties with a few in their twenties or fifties. Many were unmarried and lived alone in modest (one- or two-room) quarters, often small cabins that were intended for summer use only. Rent for such accommodations was low during the off-season when the area was cold, wet, and relatively unpopular. Alliance members sometimes traveled or changed quarters with the arrival of the summer season. As an interesting side note, virtually none of the members of the Alliance encountered in this study either owned or had easy access to a television set.

Although most Alliance members had college degrees and many had graduate degrees, individual members were slow to volunteer this kind of information; apparently believing that this aspect of their background bore little direct relation to their Alliance-related goals and activities, a few individuals with degrees even expressed mild annoyance at direct questions regarding their past academic achievements. (It is also possible that individual members wished to diffuse any possible link between the legitimacy accorded to any individual's values or beliefs and that individual's academic or other formal credentials.) In spite of their academic qualifications, members of the Alliance were almost always employed within the community (it was considered a major setback to be forced to go "over the hill" to the city for a job) and had in many cases developed highly unusual business enterprises and collections of skills in order to remain employed within their small community.

Almost all of the active members of the Alliance had established residence in the community before or around the time that the Alliance was formed but were raised in other states. Although they had lived for some years, in many cases, on very limited incomes under little better than subsistence conditions materially, they appeared to come from predominantly middle-class or upper-middle-class backgrounds. The more substantial resources of close relatives would have been available to many Alliance members in any real emergency and, in this among other respects, they did not appear to feel the full weight of their near poverty-level income and living conditions.

With this background in place, the remainder of this chapter is devoted to acquainting the reader in some detail with several individual members of the Alliance—specifically with four who attended meetings regularly ("Brad," "Bill," "Murphy," and "Art") and with three others ("Mary," "Greg," and "Peter") who were no longer or never were regular participants in the group's meetings. At this stage, my primary object continues to be to allow these people to speak for themselves. In keeping with this purpose, I have made extensive use of direct quotations. Readers anxious to move to more general understandings may, however, wish to be especially attentive to evidence of environmental commitments, concerns with a sense of community, and efforts to reconfigure work both substantively and organizationally. These themes, along with their expression in terms of energy-technology preferences, will figure prominently in the interpretive discussions offered in chapter 3.

BRAD

Brad was one of about five members of the Alliance who essentially never missed a meeting. As of 1983, he was in his fifties and lived alone in a separate single-room (plus bath) efficiency apartment at the garage end of a ten- to fifteen-year-old single-story ranch-style house. He earned his living within the community as an independent appliance repairman—a business he started two or three years earlier and continued to run from his rented apartment. His room, including a workbench (about five feet long) and hanging tools and shelves against one wall, was kept meticulously—almost exaggeratedly—neat and clean. Brad drove an old (1960 or so) Volkswagen bug, which he apparently maintained himself, with the passenger seats removed to make room for his tool chests. In 1980 or '81 he purchased a home computer that, to his pleasure, became a major continuing diversion. It occupied a prominent position on his desk and was, among other things, put to very practical use in the job of maintaining various membership, mailing, and telephone lists for the Alliance.

Never highly visible as an initiator of ambitious new projects, Brad's primary importance in the business of the Alliance was as a steady contributor to virtually all of the group's activities. In addition to his direct contributions to events, he acted for some time both as the group's treasurer and, as mentioned above, as the custodian of the group's membership, mailing, and telephone lists. As one of the group's earliest and most steadfast members, Brad was entrusted with the files kept by the group's first staff person, Frank, when the latter moved away from the community. Brad's knowledge of the group's history, therefore, as well as his extensive knowledge of the community, partly acquired through his earlier window-washing and later appliance-repair businesses, were also of great value to the group.

In order to convey an understanding of how Brad came to be where he was at the time of this study, it may be useful to follow his own discussion of his background beginning, in response to a direct question, with his academic experience. At the end of this discussion, which occurred in the course of a single interview session, Brad clarified his position in response to more specific questions, and these points also are reviewed here.

Brad holds a bachelor's degree in physics, which he got in two years at the Illinois Institute of Technology—after spending seven years "treading water," as he put it, as an undergraduate at the University of Chicago. He switched to IIT specifically so that he "could study technical things, physics, math—and do well, and then go on to graduate school." In 1960, with almost straight A's from IIT, he was admitted to and began graduate work in the physics department at the University of California at Berkeley. After finishing his course work with somewhat mixed success, he had trouble finding research that was satisfying to him. Thinking back, he notes, "My reason for studying physics was maybe not the best one: it was more [of a] philosophical" interest than anything else. "I like mathematics; I like clarity and precision, and physics is messy at the frontiers." Philosophically, he wanted to do well in quantum mechanics as this appeared to be where the answers to the problems of the universe lay at the time; but its messiness led him to spend more time and do better with the tidier classical subjects (mechanics, prequantum electrodynamics, etc.) After "baby-sitting" a professor's plasma physics experiment for a year, Brad left graduate school, although at his adviser's request he did spend another six months completing a paper related to the experiment (a paper he later regarded with some satisfaction as a respectable piece of work).

After leaving Berkeley, Brad got a job with a small research and development firm running experiments with a "magnetohydrodynamic hypervelocity gun" to simulate micrometeorite impacts on the skin of a space capsule. When the funding on this project ran out, he took a job that a friend of his was leaving in the mathematical statistics group at the telephone company. In this job he completed work his friend had begun in the development of linear programming computer codes for trunk-line allocation and generally through his work became "very valuable" to his boss and to the "math-stat" group.

At a family reunion in perhaps 1967, Brad's brother showed up with "a big bushy beard," and "he looked so jolly" that Brad himself immediately began growing his own. This "caused a lot of eyebrows to raise" at the phone company but "didn't cause any real trouble until maybe a year later when . . . my boss and I got into the elevator with the vice president in charge of our department who had a reputation at the time of being a Neanderthal man amongst conservatives."

> I don't know . . . if it was the beard per se or the fact that I didn't greet him and my boss did or . . . I'll never know what did it . . . but he got on my case and it took him, I don't know, it may have taken the better part of a year to get me out of there—to kick me out of the company. . . . He got me fired.
>
> He never did speak to me but the word came down. In fact, that was the way my boss put it, he said, '[T]he word has come down' . . . that Mr. ____, the vice president, said that 'either . . . the beard goes or he goes.' [T]hat got softened later to well, maybe he could trim it. And it wasn't an outrageous beard, you know. . . . But I wasn't willing and after a young lawyer got wind of my case and decided to take it on a, what do you call it, contingency basis, and wrote a strong letter to the management, I got fired immediately, as soon as they read the letter.

Brad was at the phone company for three or four years and got married during that period. After leaving the phone company, "I had the feeling that I didn't want anymore to work for an institution or a company and . . . that was when I first conceived of the idea of just going out on my own and doing handiwork." The economy was going into a recession at the time, however, and Brad applied for and got a job at U.C. Berkeley in the Office of Institutional Research—basically as a computer programmer. His marriage broke up shortly after his departure from the phone company.

While working at Berkeley, Brad became involved in the "food conspiracy"—a cooperative citizen program of direct bulk food purchasing from farmers—and in other activities, such as making his own bread and maintaining his own car, that were becoming popular at that time. In contrast to these "homely skills," his work at Berkeley, "[e]ven though it was fun," began to seem "like being paid, paid pretty good wages, to do crossword puzzles." Using a state university hiring freeze imposed by then-Governor Reagan, he leveraged his way from full-time to two-thirds-time then to a half-time job at Berkeley while expanding his involvement in the food conspiracy and other activities. Developing a "runaway addiction" to structuring his own time, he finally left his job altogether, after a six-month saving-up period, and launched a two-year effort to make it as a handyman. After two years he had developed enough of a business that his bank balance had stabilized and he had replaced the reserve he had begun with.

Brad originally moved to the community in which the Alliance is active with a woman he had known in Berkeley. He began a window-washing business that soon expanded to the point that he and a partner could not keep up with demand. After getting his fill of washing windows over a period of a few years, he gradually left that business to his partner to get into appliance repair. Noting that he had virtually no competition in this rural area, Brad again generated a "tremendous clientele," "simply by doing [the work] as well as I could and by being reliable and responsive."

After an initial "strong reluctance . . . to become so involved in the community through my work . . . that I would no longer be a free agent," Brad came to consider himself to be "a community fanatic." "I did get sucked in and I did become totally involved [in the community], but contrary to what I thought, I love it. [I]t's given meaning to my life."

In describing his work with the Alliance, which he joined immediately after its formation, Brad recalled having read Helen Caldicott's book, *Nuclear Madness*, after it had been recommended to him by Frank (a member, mentioned earlier, who later served as a staff person for the group). Caldicott's message was so powerful for him at that time that it triggered what he quickly determined must be a lifetime commitment.

OK, this is our work, you know, I mean the world has to be saved, and we've got to do it. 'Cause the politicians aren't doing it, you know, the power elite isn't doing it, it's got to be the people. So I made up my mind, then, that this would be a lifetime concern of mine. And I think it must be because the people who work for the opposite directions are not lying down. You know, those interests have their own motivation and they have fuel to go for their lifetimes. . . . [T]his is not something that we're going to get over with, so you better make yourself comfortable, you know? Get comfortable, and get ready to do this for the rest of your life. You know the phrase "burn out" in political activism? Well, I tell people, don't get burned out because you're not going to be worth anything. . . . It's a lifetime thing. I mean, not necessarily *the* [Alliance], but the whole issue of returning political control to the grass roots.

For Brad, "the hierarchical structure and the concentration of power at the top as it is," contributed directly to the generation of the energy, environmental, and other human predicaments addressed by the Alliance. And "I shouldn't lay it all on the doorstep of nuclear technology because even without the weapons and the radioactive wastes, we know of ways of modifying the global environment which could cause catastrophic changes in the biosphere, you know, like the greenhouse effect or acid rain." "[W]e could afford war and we could afford environmental stupidity before because we didn't have the power . . . over the physical world that we now have. . . . I believe, and many people now believe (scientists as well), that we [now] have the power to destroy life on earth; and we have built the means to do it." Under a hierarchical system, when the statistical "fluctuations are in the direction of good, that's fine, but when they're in the direction of bad, if the power is absolute, that fluctuation is completely destructive. . . . The only thing that we can do to protect ourselves from destruction by these forces is to control them, and that—I think that takes a new structure for society."

While undoing much of the cultural evolution of the ages was, in Brad's mind, a tall order, particularly in the relatively short period of time that may be available to accomplish this,

the first step . . . is for us to start to do what the hippies did . . . but with a clearer picture of what it's all about. That is, to begin to start to take power back into our own lives. You know, then it was a somewhat violent process, in fact that was at the time of a lot of violent student demonstrations, you know, [like] the free-speech movement in Berkeley. But now a new element has been introduced. Gandhi and nonviolence has come on the scene. And I think that's a crucially different thing. And we've had the example of Martin Luther King in the South of this country. So I wouldn't go so far as to say I have some hope, but you know, something like that. You know, there's a hope that maybe what was started back in the '60s in the way of taking back local autonomy can perhaps happen this time. One of the criticisms of that point of view that I've heard recently is [that] we can't go backwards. But the same thing has to be said about nuclear power and all these technologies. We're not going to unlearn those either. So, in fact, that's an even more inexorable fate than changing the political, the cultural scene. Knowledge will not be erased. But there is some chance of transforming human culture. And I think, you know, that's what we have to work on.

I feel that what our culture has grown to is an unhealthy extreme. I feel that we are infantile or childish in our having given up power. I don't mean that each of us, individually, did this in our lifetimes. We inherited the culture from our ancestors, you know, from the last generation. No one generation did this—got us where we are. It's been a growth, an organic growth. But what the growth has been is that we've given up a lot of what is being human to the power structure. We don't educate our children anymore, that's done by the educational system; we don't deal with birth and death anymore; [etc.].

I think grassroots power is important to defuse the technology, that's one thing, and also to introduce a nonviolent way of looking at the world. I think the whole hierarchical structure is basically violent. I . . . experienced it that way when I worked under it, you know, . . . being employed. It just seems to me violent to submit yourself to someone else's authority, even, no matter how benevolent the other person may be; they can't possibly know what it feels like to be you. And to give up your autonomy and do what somebody else says simply because they're the boss, I think that that's as much evil, to give up the authority, as it is to take it and use—and be the boss and order people around. I mean that's how armies are built.

This in broad terms, then, was Brad's "lifetime commitment."

There are those who might have argued that an intelligent person such as Brad was wasting his talent in a little town off in the country and that if he had really held the commitments he espoused, he would have been out there running for Congress, teaching in a university, or someplace else where he could really have made a difference. To a challenge of this nature, Brad responded:

Well, it's something that I struggle with all the time in my mind. I don't have any real struggle with it in my gut because I don't have any inclination to go to those places where power is concentrated. I fled from them, you know? I find those environments unhealthy, and I think that the people who go there who choose to be in those environments are submitting themselves to an unhealthy influence. And I'm not sure whether I approve of it, both for them personally and for the world. But I justify it in my own case in terms of what I've been telling you. I mean my value is community. If community cannot be transformed at the intimate personal level, then I think there's no hope for the world. So it doesn't behoove me, holding that point of view, to go off to the centers of power to make more power. I mean my job is to bring power back to the grass roots, and that cannot be done from the top. I mean . . . what it amounts to is giving up of pride, you know; I'm thinking right now, turning it over in my mind—I'm feeling very much like the nuns and the priests in the Christian tradition: humble. You know, I don't think it's possible to go to the top and transform the world. The transformation has to occur where you are, and where I am is with the people that I'm relating to. The people at the power—at the centers of power—are not living in community. And so they're—the very existence of their activity works against the transformation that I think the world has to experience to survive. So . . . what it amounts to is . . . I have to give up—I have to say, it's not in anybody's power to go to the top and pull this thing out of the fire. The best that each of us can do is to stay where we are and make religious community, spiritual, political, grassroots community happen everywhere. It's a humbling realization. And you know, I keep coming back to the Christian metaphor, which isn't my background at all, . . . but the Christians talk about faith and humility. Well, humility means that you realize what your place is, you're just one human being and you can't straighten out the world, that that is God's domain, to straighten out the world. All you can do is humbly serve. And faith means you have the faith that . . . by just doing that, that it will work out the way it's supposed to work out. So that's the answer for me.

[B]ut another part of the answer is, well . . . I miss the exercise of my mind that I had when I was at the university and working at the phone company. I'm getting a lot of it from the computer now [his home computer—and] I know I'm capable of that, [but] I'm not sure it's entirely healthy to let myself be led around by the nose because of that talent. In fact, I guess . . . it was Da Free John that said talent is a terrible danger. You know, you can spend your lifetime pursuing a talent but there are higher purposes for human beings. I don't want to ignore my talents either, because they're enlivening, you know; I feel very much alive when I'm getting mental exercise. Just as somebody else might feel alive when they were lifting weights or doing high jumps or something else. But still, not to lose sight of the bigger issues.

One other aspect of Brad's experience requires mention. While he was living in Berkeley and before leaving the phone company, he had joined several friends attending a number of lectures given in Sausalito by Alan Watts. Feeling that his life was "pretty much of a total failure" at the time, he was tremendously attracted by Watts' discussion of "enlightenment" and of Oriental religions and, hoping for some kind of "breakthrough," he tried LSD on three occasions. His experiences with the drug were, on the whole, intensely unpleasant and led to psychological problems that he felt for some time he could not overcome—his sense of the failure of his life, for example, became so intensified on a few occasions while not directly under the influence of the drug that he felt he would not be able to maintain his sanity. With assistance from Watts' secretary, Brad located a psychiatrist who worked with him for three years. Brad describes this as a humbling experience, one that put him "in touch with death . . . something that it's good to be in touch with." "[T]hat I could make the kinds of changes in my life that I've made partly comes from having faced that crisis and realizing that there were more important things than a lot of the things that I'd thought were important."

BILL

Another regular participant in Alliance meetings, here referred to by the name Bill, was a slightly built person whose demeanor sometimes suggested shyness. He was nevertheless a major initiator of and contributor to Alliance musical events, the most ambitious of which was *Blue Line*, the musical comedy about the 1981 Diablo Canyon blockade, which he wrote and produced locally with extensive support and participation from other Alliance members and friends. In addition to his musical activities, Bill traveled frequently to peace and antinuclear demonstrations (he was, for example, present at the 1981 Diablo blockade depicted in *Blue Line*) and acted as the principal liaison between the Alliance and several other grassroots groups and larger organizations. During and immediately after the production of *Blue Line,* he also served as a part-time staff person for the Alliance, with responsibility for keeping the office open, for taking care of correspondence, and for other functions. In his early thirties as of 1983, Bill lived on his own in a rented single-room summer cabin high on a cliff overlooking an ocean bay. This location was, much to his disappointment, only available during the winter season; he had previously rented a room in the home of a friend who often also contributed as a semi-professional singer-songwriter to Alliance musical events and fund-raising efforts.

Like Brad and most of the other active participants in Alliance activities, Bill grew up in another state, with every prospect and expectation of success in more traditional pursuits. He had, in fact, intended to become a lawyer. Some discussion of this intention and what became of it will now be provided before turning further attention to Bill's more recent work with the Alliance.

After finishing his undergraduate degree in government at Cornell University, Bill took a year off before starting law school. During that year, he signed up for VISTA (Volunteers in Service to America) but dropped out after ten days of training and instead got a job at a "really good mental hospital in western Massachusetts," the Austen Riggs Center, where he encountered what he refers to as "some really intelligent so-called schizophrenic people." During the year, he became very interested in the needs of people in this kind of situation; he had previously worked at a camp with disabled kids and was, in general, rather drawn to doing social work. Putting these interests aside, however, he started law school at Boston University.

[I]t always seemed like it would make sense for me to go to law school. I mean, all of my roommates [at college] were going to law school, my father was a lawyer and a happy man and a man who I respected, and I was pretty political. So it seemed like it would make sense.

Apparently, it did not. Among other things, he found that he was very uncomfortable with the adversarial approach itself and with the calculating and emotionless exploitation of selected facts and points of law that appeared also to be exclusively encouraged. "I mean I disliked more than two-thirds of what was happening there and from the first day I wanted to quit." "[T]here I was in law school wondering what the heck's the matter with me." He finished the first year and obtained a leave of absence, "just sort of to play it safe and to prove I could do it" but really under the assumption that he would not be going back. Returning to New York state, he instead entered the State University at Buffalo (which was cheaper, he notes, than private law school had been) and obtained a master's degree in social work.

Licensed in the state of New York, Bill then did community mental health work near the Republic Steel mill in Buffalo for a year and a half. After injuring his back in an automobile accident in the winter of 1977, however, he decided to move to an area with a milder winter. He left his job, sold his car, and traveled across the country with his younger brother (he also has one older sister), who was starting school in the state of Washington. Establishing a base in California, where he already had a few friends, he found himself increasingly attracted by the coastal communities and was soon settled in as close to the coast as he could and still be on a main bus line to the city. While looking for another job in social work, he began working in a local restaurant, unaware that he had settled just over the hill from the community in which the Alliance was active.

Pay at the restaurant, which he understands to be one of the oldest continuously operated cooperative restaurants in the country, was "terrible," and there were always "incredible" human and social problems around the place. It had been started by a "Christian," allowed no smoking at all, was strictly vegetarian, and was dedicated to providing the highest-quality food at an absolute minimum price. The combination of a young semitransient workforce and high ideals, in tension with loud rock music and fringes of the drug scene, gave the place a flavor Bill likened, with some amusement, to a Christian morality play. Although he found a community health center therapy job soon after starting work at the restaurant, he almost felt his communication and other skills as a social worker were more needed at the restaurant than on his new job. The county had many other "really intelligent, quality therapists," and the county's "clients didn't seem to be in all that bad shape," whereas the restaurant always seemed on the very edge of catastrophe. "So there was a real interesting balance between my so-called career and . . . this little job that I had had as a filler and it was—it took me over a year to quit it. . . . [I]t actually took me hurting my back again to give up the [restaurant]."

Bill held his community health center job, which happened to be not in the city but in the area in which the Alliance is active, for two and a half years. He moved to the community in which the Alliance was active during that period and characterized the end of his job at the health center as follows:

> I was there for two and a half years until I got laid off. Prop 13 [California's Proposition 13 relating to local property taxes] finally came down and so everybody without much seniority got axed. It was real ironic that the two mental health staff who lived in this community [in which they served] were the two who were axed. And then they brought more people from over the hill. It's typical. No questions of the staff as to how we wanted to deal with the budget cuts and how we would work out the staff changes—[questions that might usefully have been asked in this case] because there were other people who were ready to leave [for reasons unrelated to the budget cuts]. So what happened was, two other people resigned anyway. . . . [F]our out of six mental health staff left in one month.

With his time freed up as he left the community health center, Bill immediately spent eight days at an ongoing Diablo Canyon demonstration. He later worked again for the county, dividing his time among local school systems, but his involvement principally in nuclear power and disarmament issues absorbed an increasing amount of his energy and became, in fact, a full-time activity during the production of his musical play, *Blue Line*.

Bill first became active in the antinuclear movement shortly after the major accident at the Three Mile Island nuclear power plant in Pennsylvania. While working at the restaurant, he had become a member of an Abalone Alliance-affiliated group in that community before beginning his health center job. When members of this affiliate, an Alliance neighbor, traveled to Sacramento to call for the closing of the Rancho Seco power plant (a sister plant of the same design as Three Mile Island), he accompanied them and was "really amazed" by the whole experience. Having met some of the Alliance people in Sacramento, he joined the Alliance in the summer of 1979 when he moved to the community in which the Alliance was active. His previous ties with the neighboring Abalone affiliate and other groups in the neighboring community in part explain his usefulness as a liaison between groups in the two communities.

Before his move to the coast and shortly after his Rancho Seco trip to Sacramento, Bill organized his first musical fund-raising event for the statewide Abalone Alliance at the cooperative restaurant, which had routinely offered live music to its patrons. The event included his own first public performance of any consequence and was on the whole very successful. He subsequently organized and put on about a dozen such musical events.

> [A]long with doing the typical other kinds of letter writing and going to meetings and going to [demonstrations and] things, a lot of my focus has been musical events. To raise money and . . . inspire. . . . I've always been inspired by music and have felt inspired to sing out.

The inspiration of music is, in fact, one of several things Bill believed to be of critical importance to the success of the antinuclear, disarmament, and other social movements supported by the Alliance. Other critical needs to which he was attentive included the need for "heroes—people's stories that can be identified with" and the development of a more "universal positive vision of . . . how we can succeed by working together."

At a 1983 disarmament rally attended by Bill Perry, the Lawrence Livermore Laboratory public relations executive who resigned to join the California Nuclear Freeze campaign, Bill combined his musical talents with his sense of a need for heroes and quickly produced the following song, which he sang at the rally:

Verse 1

Well, William Perry's my name
  And I worked for the weapons makers.
'Til all those protesters came
  Acting like givers not takers.

In the late spring of '82, I was troubled,
  Not sure what to do.
I found my bosses had led me on
  And I was working for men who were going wrong.

Chorus

The day I opened up my eyes
  To see my brothers and sisters.
The day I saw through all the lies
  To see those brave resisters
They were standing there arm in arm
  To save our whole world family from deadly harm.

Verse 2

I was a slick PR man
  I knew how to play with words.
But I'll tell you, selling nuclear war
  Is strictly for the birds.
Well now a fat paycheck can make a man feel good
  Like a reward for doing what he should
But when my Livermore cronies said, "Way to go,"
  A voice deep in my heart was 'a cryin' no!

Chorus

It was the day I opened up my eyes
  To see my brothers and sisters.
The day I saw through all the lies
  To see those brave resisters.
You were standin' there arm in arm
  To save our whole world family from deadly harm.

Verse 3

Well, like Dan Ellsberg before me,
  I got to raise my voice.
More than any time before,
  We all face a crucial choice
Well, we can hide from the truth
  With drugs and TV
Run in fear from Helen Caldicott's plea
  Or we can join together to do our part

And each in our own way,
  Follow our hearts.

Chorus

Today I opened up my eyes
  To see my brothers and sisters.
Today I saw through all the lies
  To see you brave resisters.
You were standing there arm in arm
  To save our whole world family from deadly harm.

Perry had come to the rally at Bill's invitation and the new song, sung to the tune of "The Night They Drove Old Dixie Down" (Robby Robertson), while it was not expected to replace any classics, was apparently well received.

Bill's principal Alliance-related effort, the musical comedy *Blue Line,* can also be seen to flow logically from clearly articulated objectives. Among these objectives was the desire to contribute to a more "universal positive vision of . . . how we can succeed by working together." In outline, the play follows the exploits of a group from the fictitious town of Esperanza, as its members travel to Diablo Canyon and participate in the 1981 blockade there. The title refers to the blue line painted on the pavement at the power plant's entrance, separating public from private property—crossing the line made one subject to arrest—and the main character in the play is autobiographically inspired. Lasting about two hours, the play contains many original as well as many well-known songs and provides an apparently accurate and detailed picture of the process of participating in the event. It offers realistic portrayals of the practice of organization by affinity groups (the Esperanzans identify their group as Vaya Con Dios) and of the consensus decision-making process. It depicts the full course of the demonstration process from nonviolence training and the nonviolence code all Abalone-affiliated blockaders were required to sign before they could participate to the actual arrest of protesters at the blue line. Throughout its course, the play pokes fun at both sides. A superconservative state trooper, for example, sings in one amusing scene at the blue line about seeing "aprons and pigtails" on some of the male demonstrators, and about how some of them "eat the weirdest stuff"; dedicated to "the way it ought to be," the trooper "can't wait" to get his hands on them and haul them off to jail. The play is also filled with information and more serious messages. In the same scene at the blue line, another state trooper doubts that the money spent on the reactor was any better spent than the money being wasted paying state troopers to arrest demonstrators; he points out to a reporter, with whom the troopers are conversing, that there are "many different people wearing these uniforms" and the reporter responds, "[N]ice to know—it might be harder to tell when you're all dragging people off to jail for crossing a blue line."

Blue Line was videotaped at one performance and there was some hope that it might be shown on at least local educational and/or cable TV channels.

> I'd like it to really succeed. And have more mainstream people be interested in it as a play and a musical and then come to discover that the people who are involved in the antinuclear movement are people. Ordinary kinds of people who have different kinds of personalities and being alive at this time in the world, feel this commitment, and try to fit that political commitment into their lives . . . and just how, you know, a really unusual event like a blockade fits into people's ongoing lives.

Bill also hoped to convey his sense that the 1981 blockade and associated resistance efforts were effective—that they helped draw enough attention to safety and other issues to be an important factor in the discovery of design flaws in the plant's earthquake features and, hence, in the decision to delay substantially the opening of the plant. The chorus of what might be called the theme song of the play runs as follows:

> You can make a difference in this world.
> You can be the way you want to be.
> You know what you feel inside.
> You know what you can see.
> And whatever's comin' down you can be free.

The play closes with the full cast singing this chorus together.

Although he was not explicit about it and did not attempt to develop it in *Blue Line*, there is an interesting hint of what might inadequately be described as a kind of religious faith or conviction in Bill's discussion of the actual 1981 Diablo Canyon blockade. In describing his own experience with the blockade, Bill referred to

> the whole magical way the timing [worked out], finding out about the design flaws the day after the blockade ended. . . . I mean it was like this act of God. I mean it was really incredible. . . . [T]hese people did a blockade, you know, having no idea that they would have impact, hoping they would, and it worked. At some level. Because you know, Diablo would have been radioactive two and a half years ago if we hadn't done that blockade. . . . [I]t's just true. And so, you know, we can make a difference, even if it's unclear how it all fits in. . . . [H]ow the work that we're doing with one another fits in. . . . [I]t's my belief that it does. And that we are doing the task of keeping life on this planet viable.

Believing that "everybody who is aware of dangers has some responsibility to do whatever they feel they need to do," fitting whatever actions they decide to take "within their own personality and their own lives," Bill wrote *Blue Line* "to underline that historical time" in which collective exercise of that personal responsibility effectively changed the course of events, in his view.

In addition to his musical activities and frequent participation in "political actions," Bill was one of the Alliance's most active individuals in a more traditional political sense. His efforts to maintain contact and cooperation with neighboring groups were complemented by personal contacts in Washington state and elsewhere and by memberships in the Fellowship of Reconciliation and other major organizations. He wrote regularly to his political representatives, to Nuclear Regulatory Commissioners, and to others in positions of influence in the areas of concern to him, corresponding perhaps most frequently with the area's congressional representative. He and other Alliance members also met personally with this particular representative to discuss specific issues of mutual concern. In addition, Bill made extensive use of the newspapers—the editorial pages, news reporting, and advertising—often to publicize Alliance concerns or events. An extensive news feature, for example, which he wrote about the Diablo Blockade, was almost published in *East/West Journal* (they held on to it, expressing a strong interest for three months but finally printed another account); it was printed in condensed form in the local paper and excerpts were accepted for publication in a book about Diablo Canyon.

All of these activities combine in an interesting fashion with the fact that Bill had (as of 1984) not yet chosen civil disobedience and arrest as a further means of expressing his convictions:

> [I] have never really wanted to get arrested as yet. I kind of imagine that partly it has to do with my dad being a lawyer. I've always [taken] civil disobedience real seriously as something you do when you have to do it, when you've tried everything else. And so I'm trying everything else.

MURPHY

The third regular participant in Alliance meetings to be described here will be referred to as "Murphy." Unlike most of the other active members of the Alliance, Murphy lived in the community well before the Alliance was formed and did not have a college education. He earned his living principally as an auto mechanic and in home repairs and remodeling. From his arrival, he pursued both of these trades almost entirely within the community. More recently, he also worked periodically with another active Alliance member (to be described in the next section) on home weatherization for energy conservation, again almost entirely within the community. In addition to his Alliance activities, which included playing the lead role in *Blue Line*, Murphy was very active in a separate local organization attempting to establish bicycle paths in the area and in a small West Coast religious group of modern origins.

Before describing Murphy's Alliance-related activities and attitudes in any greater detail, a review of some of his earlier background may be useful.

Born in 1949, Murphy was the youngest of ten children born over a twelve-year period. His father, a medical doctor in general practice, and his mother, a University of Utah graduate, were both once-active members of the Mormon Church. His oldest brother became a state president in the Mormon Church, but Murphy himself, his father, and other siblings eventually withdrew from active participation in the church. He spent the first eleven years of his life in an upper-middle-class suburb of Denver, Colorado, but when his parents separated, he moved to California with his mother. He, his mother, three brothers, and one sister then lived in the city near the Alliance area (initially in the back of the hat shop in which his mother worked) until, as he reports it, each of the children got old enough to leave and not be brought back.

Murphy left home six months before he would have graduated from high school and began working as a car mechanic. He married, and he and his wife had one daughter shortly after his eighteenth birthday. After three years of marriage, his wife left him, initially taking their daughter with her. Six months later, however, she returned their daughter to Murphy, who cared for her between that time and the time she entered her teens, except for visits to her mother for a week or two each year.

Murphy used the term "culture shock" in describing his move from Colorado to a city environment in California. (The departure from Colorado actually occurred on his eleventh birthday.) Green grass had suddenly been replaced by beatnik coffee shops across the street and "strip joints" around the corner. While still living with his mother and perhaps partly seeking a more familiar environment, Murphy on more than one occasion rode his bicycle on the rather lengthy journey from the city to the area in which the Alliance would later be formed. After marrying, he and his wife also drove out repeatedly in a concerted though at the time unsuccessful effort to find a place to live there in the country. Around the time of his separation from his first wife (around 1970), Murphy did, in fact, move into an old school bus in one of the three towns in which the Alliance has since become active. A pot-smoking, aspiring guitarist at the time, with shoulder-length hair, he had dreams of touring the country with a rock band in the old bus.

As Murphy remembered it, the dream of traveling with a rock band was, for him, closely connected with two contrasting images of the time: that of President Kennedy's assassination and that of the Beatles. The assassination of John F. Kennedy—an event that occurred while Murphy was in junior high school—was a vivid memory for him and he drew a link between it and the whole anti-Vietnam War "why stay in school?" ethos later prevalent among young people during his high school years. He accounted for some of the popularity of the Beatles around that time by noting that they appeared to offer a kind of happiness in life that was uniquely enticing in its sharp contrast to the more negative currents of the time. As things worked out in Murphy's case, of course, there were no national tours. He lived with his daughter in the bus for two or three years and, in 1983, described the experience as "an interesting way to live" but noted that "it wasn't very comfortable."

Murphy later married again and, staying in the same community, moved to an old schoolhouse with his new wife, her two daughters from a previous marriage, and his own daughter. Five years later, this second marriage also broke up. Murphy then married a third time. His third wife also attended Alliance meetings and had lived in the community for some time; the two first met, in fact, at one of Murphy's first Alliance meetings in 1980.

Between his second and third marriages, Murphy lived with his daughter in a small two-room cabin in an isolated pasture across the town road from the edge of a small ocean bay. An outhouse near the cabin served in the absence of any sewer connection or septic field, and kerosene lamps provided the only artificial light (the cabin had no electric service connection). Murphy still rented this cabin in 1984, although his wife also had a small summer cabin (which *is* connected for sewer and electric services) a few miles away; the family spent about half the time in each place. (Murphy had also retained his old school bus until his wife pressed him to get rid of it in 1982.)

Well dressed and well groomed at the time of this study, Murphy gave no clue through his appearance of the unconventional way in which he had lived in earlier years. Soft-spoken and always well organized and to the point when he did speak, he had all the attributes that would, in fact, lead one to place him among the successful young urban professionals who might have had a second home or parents with a second home in the Alliance community. Interestingly, his chief regret concerning his decision to become a mechanic before finishing high school was that, as a mechanic, he found that he did not have as much freedom to choose the people he spent his time with as he came to feel he would have had if he had gone to college and on into other kinds of work. While he did not draw the link explicitly himself, one might suspect that this regret contributed to his move to the Alliance community, to his activity in the Alliance and other community groups, and to his diverse employment as a mechanic and home remodeler, etc., within the community rather than as a conventional shop mechanic in a more densely populated area.

At the conclusion of the observation period for this study, Murphy decided to get some training in the use of small computers and to explore the possibility of changing his income-producing career. He had worked with two friends (Brad was one and a less-active member of the Alliance was the other) as they got started with their own home computers, and more recently he and his wife had purchased their own small machine. Murphy noted that while he had had trouble with English courses in school, he had always liked numbers and done well in mathematical/analytical courses in school; these earlier indicators added to his sense that he might do well in computer work.

Murphy's income and expenses were, by normal standards, exceedingly modest, and a large fraction of his time was spent in activities that did not directly contribute to his income. In addition to his family and Alliance-related activities, he devoted much of his time to a local religious group and a "community paths" organization attempting to establish bicycle paths in the area. The religious group or "spiritual path" in which Murphy had been active over a period of five or six years was small, modern in origin, and not widely known. It relied heavily, however, on basic Judeo-Christian teachings and had, among other benefits, been helpful to Murphy in overcoming long-standing feelings of resentment toward his mother. During the course of this study, the community paths group, with two hundred members and an eight-hundred-name mailing list, made a fifty-thousand-dollar proposal to a private foundation for a detailed design study of two specific bicycle path options in the community, one of which would have utilized an old railroad right-of-way at a safe distance from existing streets and car and truck traffic. Cycling was popular in the area during the summer, and increased traffic, combined with narrow shoulders on existing roads, contributed to dangerous cycling conditions. If the bicycle path study was funded, Murphy intended to apply for a quarter-time position with the project and take responsibility for community liaison.

Murphy's first real participation in Alliance activities came after meeting "Art," another active Alliance member (who is introduced in more detail in the next section) at a meeting of the community paths organization. Art was organizing the solar fair put on by the Alliance in 1980, and at his suggestion, Murphy began attending Alliance meetings as a way to get familiar with and begin contributing to the solar fair project. From 1980 to 1984 he was an active participant in and contributor to the group, with a continuing special interest in alternative energy systems. "I'm not anti-anything much," he once said, although he was very supportive of other Alliance efforts and of other members of the group. (As noted earlier, he did perform in the lead role in *Blue Line,* a major commitment in time and energy to a project that was surely more directly antinuclear than prorenewable. It might be more precise, therefore, to say that the activities he initiated or advocated tended to be more prorenewable or probicycle paths, etc., than anti- anything). He was a strong advocate of wind energy systems in the community, which is often swept by stiff ocean breezes. And, when he could, he enthusiastically assisted the same "Art" who initiated the solar fair in the latter's more recent home energy conservation contracting business. As yet another example, he was interested in a photovoltaic system to replace or augment the kerosene lamps in his home and was one of the principal advocates for such a system at the Alliance office as well. In addition to his regular attendance and contributions to decision making at Alliance meetings, Murphy was a regular contributor to the organization and execution of Alliance-sponsored community events and could also be counted on to lend his skills as a guitarist and vocalist whenever music was required.

With reference to his lack of higher education, his work with cars and home remodeling, and his more recent decision to explore career possibilities in the computer business, Murphy had, as of 1984, done some reflecting on the lives of people he had known who operated principally in the world of ideas and had little "physical" (practical or mechanical) experience; he felt that his own natural tendencies would ordinarily have made him particularly vulnerable to the hazards of a life of ideas and considered himself in many ways fortunate to have had other kinds of experience. However this may be, his attitudes and actions tended to be less "studied," less intellectualized, less consciously articulated than those of other Alliance members who attended meetings regularly. His participation in Alliance activities had, in fact, greatly contributed to his ability to articulate positions and concerns with which he apparently identified. He referred, for example, to the way in which information that was new to him on uranium mining, on the cancer effects of radioactive materials, and on nuclear wastes, got him "into consideration of what's right to do with nuclear" power. An independently produced film on energy resource development in the Four Corners area that was shown at an Alliance-sponsored community event also appeared to have had an impact. (Among other things the film notes official government statements that the area may have to be designated a "national sacrifice area" because of unmanageable environmental impacts.) Yet at a common-sense level, Murphy gave expression to the same sorts of skepticism regarding current practices and to the same desires for what he believed to be better ways of doing things that underlie more sophisticated articulations of the same positions.

> I kind of see that we're not doing it [solar and renewable energy systems generally, and photovoltaics in particular] because of the same reason why they made everybody buy black-and-white TVs before they really brought out a lot of color TVs. They had them and . . . the word was out that they had color TVs, but they didn't really push them and they kept the price real high and [in the end] they started selling black and whites real cheap.

"In the system we work under," he argued, the same thing has also been done with eight-track tape machines, videotapes, and computers. In a second common-sense reaction, referring initially to microwave hazards, he once noted:

> [T]hey say, "Oh, don't worry about it—there's not enough that you have to worry about it." But I just wonder about that. . . . You know, when I used to drive under . . . power transmission wires and the radio would always go [he imitates crackling noises of radio static]—I mean you can't tell me there's nothing going on there [he says this with a laugh].
> It's the opposite of that living in [my cabin on the bay], where there's no electricity at all. . . . [I]t has a different feel, somehow.

In a general sense, Murphy appeared to have arrived at much the same positions that other members of the Alliance had adopted, in a studied fashion, by voting with his feet. There can in any case be no doubt that Murphy had found a congenial home for his attitudes and beliefs within the Alliance:

> [T]he [Alliance] is really great because it's where people with like minds and like ideas can get together and feel supported. It's really wonderful. It's a real nice family-type thing. [My wife] and I pretty much met at those [Alliance] meetings and we feel really close to those people.

Art

The last regular participant in Alliance meetings to be described in detail here moved to the area from Evanston, Illinois, early in 1978 and joined the group at the time of its formation. In his late twenties, "Art" (as he will be known here) earned his living from 1978 to 1984 through an assortment of activities that included work as a caretaker on a large piece of property in the area, full- and part-time teaching in local and neighboring schools, and odd jobs at a local "bed and breakfast" inn. By 1984, however, an increasing fraction of his time had begun to be devoted to his own small business as a home energy conservation contractor; he had been trained and was the only home energy contractor in the area who was certified by the electric utility to perform services under its low-interest home-energy conservation loan program. In addition to his regular participation at meetings and in Alliance-sponsored events, Art was the initiator and principal organizer for the fairly elaborate solar fair put on by the Alliance in 1980. He also made contributions to, or was directly active in, virtually every other local organization in the area that was consistent with his attitudes and beliefs—including the bicycle path group mentioned in the previous section, a local environmental action committee, the local American Civil Liberties Union chapter, and others. Art through his activities in the community, like Bill, Brad, and other active Alliance members, had come to know a surprising fraction of the community's residents; when in the course of conducting the observations for this study, the author sought a place to live in the community, Art was able to provide details on virtually every space advertised in the local paper as well as information on nearly as many equally good prospects that were not advertised.

Art came to California shortly after completing his bachelor's of fine arts degree at the Art Institute of Chicago. Strongly influenced in his high school years by Buckminster Fuller's *Operating Manual for Spaceship Earth*, and other popular works of the time, he came to California partly out of a long-standing desire "to be closer to nature." "As nice of a city as Chicago is—and I still like it and I go back there quite a bit. . . . It just . . . wasn't a real nice place to live." A brother had moved to one of the smaller cities near the Alliance area and Art drove west in a van, arriving in time to join his whole family for a Christmas 1977 reunion at his brother's home.

Using his brother's home as a kind of base but living to a large extent in his van, Art arrived in the Alliance community mostly by chance. He had a teaching certificate and was looking for work as an art teacher but in the meantime, of course, had no income. Acquaintances of an old college roommate of his had suggested that he apply for food stamps to help tide him over and, in the course of investigating that possibility, he was told by social-service officials that the Alliance community was probably one that would not seriously object to having him around living in his van. He came out to explore the area and, while looking in the window of a small community center on his first day, he happened to encounter the town librarian, who in turn guided him to a small social-service office less than a block away. At twenty-two years of age, he was told, he still qualified for a minimum-wage job with the Young Adult Conservation Corps in the park nearby. While still in high school, he had greatly enjoyed six weeks of work as a volunteer in Olympic National Park with the Student Conservation Association ("It was one of the high points of my life"), so he was happy to begin working at the park and with equally good fortune was soon settled in a small rented "in-law" cabin of his own.

Art's involvement with the Alliance began almost immediately after his arrival in the community.

I had wanted to work in the antinuclear movement from the time I lived in Chicago. However, in Chicago there were very, very few people involved in the antinuclear movement—it was really heavily,

heavily, heavily nuclear. And there was very little concern. . . . [O]ne of the reasons why I came to California was because I wanted to escape that heavy nuclear mentality from Chicago. [And as soon as meetings, nonviolence training sessions, etc.] started happening in a place that was easy for me to get to, I started to go to them. . . . and that was . . . probably February to March of '78.

Art's participation in the Alliance continued without interruption. His employment with the Young Adult Conservation Corps, however, was only temporary.

Fortunately, with the help of his boss at the park, Art was able to move directly from his temporary park job to a Comprehensive Employment and Training Act job as a caretaker at a military base some distance away. Not above a certain amount of mischief as a means of stirring staid conventions, he managed to have some fun in this job. The base itself, as it turns out, was being considered for possible closure as an unnecessary expenditure of federal funds, and a variety of proposals for its conversion to nonmilitary use had been made. One of these proposals suggested developing the site as a kind of proving ground for new energy and environmentally sensitive approaches to life in a residential community. Personnel at the base were generally unenthusiastic about this proposal, and Art, who felt differently about the matter, still derives substantial amusement from the reactions he got to a bumper sticker that he put on his car in support of the proposal while he was still employed at the base. On one occasion he was picking up paychecks for his group at county offices and encountered public hearings on the future of the base that happened to be under way. He briefly offered his own testimony on the wildlife and on the natural beauty of the area in which the base was located. Major areas of land at the base apparently still remained undeveloped. As it turned out, the base commander had also been at the hearings and, as Art reported it,

I heard about it. [Interviewer: Did you really?] Oh yeah. He [the base commander] was mad as an SOB. He was really—he was ready to fire me. But my supervisor, he knew that I was a good worker and he saved my job for me. You know, he said, you can't fire the guy, you know, I'll tell him he can't go to the [county building] anymore.

Since Art had no intention of developing a career at the base in the long run, he was able to take all of these reactions to his own attitudes primarily as a source of amusement.

Immediately after his arrival in California and before he came to the area in which the Alliance was active, Art had applied for work as an art teacher in several neighboring communities. He eventually did obtain a temporary teaching position, which he held for two years in one of these communities. More recently, he again taught part-time in the public schools in the Alliance community itself, although with the development of his energy conservation business, he appears to have stopped seeking work as an art teacher.

It was while he was working in the second of these teaching jobs and while drawing a partial unemployment benefit (the second teaching job paid far less than the first) that Art got the idea of putting on a solar fair. He had attended a training conference sponsored by the American Friends Service Committee nearby, and was much taken by the suggestion (made by trainers from the Movement for a New Society, located in Philadelphia) that the antinuclear movement was in the process of becoming "the people's energy movement." According to his report, it was argued at the conference that "we've really won the battle against nuclear power" and suggested that "we really have to take it a step further now and make it a battle for energy conservation and solar and renewables." It was further argued, according to Art,

> that this is a very important turning point in the antinuclear movement. To take the network, the grass roots that we've already got established and say now let's use that same grass roots for safe energy, you know, for alternatives. Well, that idea didn't go over so big [at first] with [the Alliance], but it hit me very hard.

Teaming up initially with his girlfriend, who was also very enthusiastic, and with musicians he had met in an organization supporting conversion of the military base where he had worked, Art set about organizing a major community "safe energy" event. Initial plans for a concert to be provided by Art's musician friends slowly evolved and, when the site for the concert fell through and the original musicians dropped out, the event eventually became the solar fair held on vacant land adjacent to the Alliance office in 1980.

As Art recalled it, organizing the solar fair had quickly become "a full-time job," although he did continue with his part-time teaching job during this period. He and his girlfriend began by writing letters to everyone they felt might be interested in participating or contributing. In addition, they got letters in the local paper both to advertise the event and to call for assistance, locally. (One local volunteer, who turned out to be a young architect, phoned Art "out of the blue" and ended up doing drawings, displayed at the fair, of a solar greenhouse addition to the Alliance office; these were later used in the group's actual construction of the office greenhouse addition.) A very energetic person, Art was able to combine his efforts with support and assistance from friends in the Alliance, from members of the community generally, and from outside exhibitors to produce an event that was, in Art's view, "a tremendous success in terms of educating people." Participants in the fair included representatives from *Coevolution Quarterly* and a private research institute with ties to the University of California; three photovoltaic manufacturers were represented and a full-scale wind machine was on display on a ten-foot tower; a nonprofit public interest organization was there and "did their whole tent display with their food driers" and other devices; "we cooked a soy burger on a solar parabolic cooker;" "we had . . . solar ovens" and everything else, it seems, that one could hope for in a community solar fair. Despite fundraising efforts before and during the fair (exhibitors were asked, for example, for fifteen dollars for their booth space—"some paid, some didn't pay"), Art personally ended up losing about three hundred dollars on the whole effort. He clearly had no regrets, however. While somewhat surprised in retrospect that all of this (including his own three hundred dollars loss) originated "just from my idea of, well, let's have a little solar fair," he recognized that he could really only blame one of his own tendencies: "I mean when I go to do something, I [go] all out to make it a big success." Besides, "I got a good slide show from it" (he preserved it all on film).

As his work on the solar fair should suggest, Art was an active, at times almost restless person, with definite innovative/creative inclinations. Often seemingly half-serious and half-amused by whatever he was doing, he was not generally inclined to be overtly philosophical. His unique style was perhaps most apparent in the pleasure he seemed to derive from his capacity to find unusual and attractive places to live and work. After the solar fair, for example, he managed to get a job as a caretaker on a very large piece of land owned by a private foundation. He also managed to work out an arrangement that allowed him to live on the property in a large, unoccupied house. Although there were accommodations on the land for a large number of people, these were unoccupied during the months of his employment, so he had the whole property almost entirely to himself. The structure he lived in overlooked the bay and had originally been the eastern terminus for one of the first telegraph lines to be run under the Pacific Ocean. Clearly pleased with this enviable setting, Art also enjoyed showing occasional visitors through his private art gallery—a collection of his own original works carefully displayed in one of the large high-ceilinged rooms of the venerable structure he inhabited. At another time during his stay in the community, Art occupied a circular room at the top of a turretlike tower at one end of a local bed-and-breakfast inn. The brainchild of its owner, a patent lawyer from the city, this inn began as a private summer place when a very large tree had to be felled on a steep slope on the land and the owner decided to mill the wood and build something rather than burn it or have it hauled away. The tower, which emerged as part of the project's continuing evolution, consisted of a single room with windows all the way around its perimeter resting at the top of a several-story column of smaller diameter that enclosed a circular stair. The deck on the flat roof of this room was about eighty feet from the ground on one side. Although unusual looking, the inn as a whole was solidly constructed and very attractive with massive, rough-finished wooden beams, copper gutters and trim, brick walks, and other appointments. Art worked on this structure periodically for several years and was able to participate to some degree in the generation and refinement of ideas that were incorporated in its informal and organic development.

Despite his tendency to appear more active than philosophical, Art had clearly done some thinking, and his activities were far from random. This observation finds support, for example, in his reactions to the direct question, "What's wrong with nuclear power?" While his immediate response was, "That's kind of a silly question to ask," he then continued,

I'm trying to think of something else that would be more appropriate along the lines of [a previous question concerning what] I do for a living. And that is that I think it's . . . important for you to . . . know that I think that there's a certain level of meditation that . . . is important. And I think that it's become very significant, in my life anyway, that I've begun to do a daily meditation, and I think that, as more people begin to do that . . . we begin to change our pace and change our attitude and . . . change our "energy." . . . [T]hat's maybe not something I do for a living, because I don't get paid for it, but it's something that I do as I'm living—meditation and Yoga.

In another brief but perhaps revealing exchange, Art offered the following view of the Alliance and his role in it:

[The Alliance's] objectives are to educate the community that we live in and to educate the community about safe energy. And, in my mind that's very, very broad. That doesn't just talk about appropriate technology, that doesn't just talk about the dangers of nuclear power, . . . it also talks about personal energy, psychic energy, you know what I mean? The Gandhian philosophy of energy, you know; Gandhi was a man of tremendous energy [laughs]. So in my mind, it has a whole other level of safe energy that nobody [talks about specifically]— that's, you know, California energy—[California] definition of the word [laughs]. [Jokingly:] Oh, my "energy's" really low today, you know? There're some people who actually have a dangerous energy about them.
(Interviewer: "A dangerous energy about them?")
Well, yes, and we have a safe energy about our group, I like to think. And also, when I talk about educating the community, that includes ourselves as well, because we're definitely a part of the community. My objectives as a member of [the Alliance] are to facilitate that happening—. . . and to get my input into it—to put in my two cents about what I think educating the community about safe energy means, and what I think is a good way to do it. . . . [Participation in the Alliance] also gives me a chance to express my own concern. . . . [I]t's a way for me to demonstrate that I care.

When Art was in high school, he took a class on Jose Silva's "mind control" that left a lasting impression on him. Continuing to make a conscious effort to "keep [his] thoughts positive thoughts," he credited some of his success in finding employment and in establishing himself in the community, often in very interesting settings, to this effort. "Which I guess is another reason I think what's wrong with nuclear is a ridiculous question. I'd rather say what's right with what I'm doing than what's wrong with what I'm not doing."

What he was doing in 1984 as a principal wage-earning adjunct to his participation in the Alliance and other community organizations was to develop his home energy conservation business. This business had gradually become sufficiently profitable to displace other part-time activities and to justify the purchase of a new, imported pickup truck.

> It seems to me like it always feels good to leave a place in better condition than when you found it . . . that's the kind of thing I'm talking about. And that's why when I go into a house, I want to try and leave that shelter . . . [i]n as good a condition as it can turn out. Because that's something that, also, I don't feel like anyone else can do. These insulation contractors can't do that. Because they have, you know [people] that they send out with the blow-in trucks to just blow the insulation into the attic and get the hell out of there to the next job. We can go into the house and do everything that it takes to tighten up the house. So, I think that's important. And then, after we've done that we can go back to the house later on and say, well now we know if you put in a greenhouse in here, you're going to get some performance out of it. Because we know you got all your outlet and switch gaskets in, and all your penetrations and bypasses through the floors and walls and ceiling are caulked up and . . . [We are able to] take a look at the whole situation and put it together properly.

Less Active Members: Mary, Greg, and Peter

With this description of the background and activities of four active members of the Alliance in place, attention is now turned in somewhat less detail to three less active members. (The term "less active," again, simply implies that these members did not attend the group's meetings regularly during the study period.) Unique commitments and lives patterned around those commitments will be as much in evidence here as among more active members.

Mary

Mary was born and raised in Illinois and went to a small lib-eral-arts college—a free Methodist school—in the southern part of the same state. After graduating with a major in English, a minor in religion, and a teaching certificate, she worked for a year as a secretary in a junior high school (also in Illinois), hoping that a teaching position would open up. With no teaching prospects apparent, however, and with the feeling that she was at a point in her life when she was relatively free to do whatever she might want to do for a while, she moved to a small town in Arizona and took up a furniture-making apprenticeship. (She had previously visited a friend there who was apprenticed to the same furniture maker.) Although the cabinet business she had joined closed down entirely only three months after her arrival, Mary was al-ready becoming very attached to the Arizona terrain, which she later continued to find much more appealing than that in either California or her home state of Illinois. She found a job as a fire inspector and fire safety educator in the town and settled in hap-pily for three or four years.

> [But then] they had budget cutbacks in the city and the chief elimi-nated the fire prevention bureau—which was a stupid move and just totally political, but he did it, and my job, therefore, got eliminated.

Unemployed, Mary went to California to spend a week visiting another friend who had recently moved to one of the communities in which the Alliance was active. Again attracted by the area, and again feeling the need for a change after she had returned to Ari-zona only to languish for six months on unemployment, she moved west.

Arriving in the community in 1980, Mary taught part-time for nine months while she also worked part-time in a local grocery store and occasionally earned small sums playing her cello. (The second of these jobs, she noted, provided her with a broad intro-duction to the people in the community.) More recently she lived, worked, and acquired part ownership in a local bed-and-breakfast inn (not the one Art was involved with).

Raised "in a very conservative, fundamentalist, Bible church" environment, Mary reported that she was really inactive politically until the late 1970s.

> [Even] during the Vietnam War I just stayed sort of oblivious to what
> was going [on—partly] because of the structure of my life and [the
> fact that] the college that I went to was relatively untouched and isolated.

Mary was, in her own words, "very naive and uninformed" during that period of her life but also perhaps "overwhelmed" by the events of that time. For many years, however, she was a reader of *Sojourner's* magazine—"a radical Christian magazine published by people who stand for peace and justice to the poor and oppressed." And in the late 1970s a *Sojourner's* article about a group of people practicing "war tax resistance" caught her eye. For four years after reading this article she planned to take action herself and withhold the fraction of her income taxes that were destined for expenditure by the military, and in each of these years she felt guilty as she relented and instead paid the full legal amount. Finally, she did act on her plans. Taking an illegal war tax credit on her forms, she applied for and received a refund on taxes that had been withheld in her name; instead of leaving what amounted to 50 percent of her tax liability (an estimate of the fraction of the federal budget that was allocated to military spending) with the federal government, she sent it to a nonprofit organization that in turn distributed the money to other causes that, in her words, "I feel are what the money should be going to." While it should be noted that the nonprofit organization involved here undertook to return contributions to anyone caught by the Internal Revenue Service, Mary was quite aware that in breaking the law the back taxes she might have to pay at some point in the future could be the least of her worries. "And obviously it's only symbolic," she said. "I mean, it's not that much money. It's just that my conscience would not allow me not to do it another year, regardless of what the cost was going to be."

Mary's first real contact with the Alliance did not occur until after she had carried out her plan to become a war tax resister.

> I had known for a long time that these people surely would be people
> that thought just the way I did, but [I didn't] want to get stuck, you

know? [I assumed that] if you let them know that you think like that, they're going to make you do things all the time. They're not ever going to let you go, you know, . . . they're going to go, all right, you want to get up on stage and tell people and go on this rally and that, and so I just never let them know. And that was really the reason why I didn't get involved with them.

After filing her forms as a war tax resister, however, Mary happened to get in touch with Brad (who is widely known in the community as an active member of the Alliance) regarding a war tax resistance demonstration in the city. She had decided to go to the demonstration and hoped to locate someone else interested in going with her. Brad gave her several names, and another active member (not described in these pages) did attend the demonstration with her. But Brad also suggested that she come to an Alliance meeting at least to explain what she was doing, arguing that the group could give her a lot of support, maybe even "concrete support when the chips are down [but] at least . . . emotional support."

> So I thought, oh yeah, I guess I could do that. So I did and found out that they were very low-key. [Brad had] said, "You don't ever have to come again," you know, [and] it was real nice. It was real important because of how I felt, and I would like to let the rest of the community know that's true about [the Alliance] because there must be other people like me that are afraid of the same kind of thing.

After her first meeting, Mary did not indeed become a regular participant in Alliance meetings and was only rarely active in Alliance events. While she stated rather categorically,

> I can agree with these people, yes. There isn't anything they stand for that I disagree with.

she also clearly chose the action she wished to take and did not deviate from that choice.

Some of Mary's commitment may have stemmed from her despair over her mother's attitudes as she was regularly exposed to them in letters and telephone conversations.

> One thing that she said was, "Don't you know there isn't going to be peace until Jesus comes again?" Which means, why even try to make

it happen—which to me is just heartbreaking. . . . [A]nd when I think how many, many people think just like she does, I mean . . . maybe the "moral majority" isn't really a majority, but it's a lot of people, you know? And I don't think they think. It's like the way [my mother] doesn't really think for herself. Everything she's said to me [has] sounded like quotes. It didn't sound like anything that really came from real thought.

Mary seemed to some degree to associate her mother's attitudes with a way of life that was remote from the influences of the city. Even where "there may be people [in remote areas] who think like me," who are more "awake" than Mary felt her mother was, "they're not doing anything, you know?"

Yet Mary was not entirely happy in California either. Referring partly to her experience working in the bed-and-breakfast inn, which had a city clientele for the most part, she noted that a relative heightening in "people's level of awareness" too often seemed to be accompanied by a "sophistication or pseudosophistication and class consciousness—and [a desire for] nice things all the time—that's very wearing on me." This concern did not affect her feelings for members of the Alliance and the immediate community.

[T]he way it's possible to connect to people here, I don't expect to find very easily anywhere else. I think it's real special.

Yet the "sophistication" or "class consciousness" she occasionally encountered produced a discomfort that sometimes made her think of moving back to Arizona.

Talking with her after she had returned from a visit with her sister and her sister's family in Missouri, Mary had this to say about the pattern of life in which she was raised and about the contrasting pattern she had chosen more recently:

I don't want to be that way [the old way] anymore . . . and yet I do, you know? . . . [T]here's this real strong feeling that I want to just live. Live my life, you know? I don't want to do war tax resistance. I don't want to be part of [the Alliance]. I just want to have a house and, you know, either a family or . . . a community family or . . . just let me live, you know? And that's probably part of why I didn't get involved [in the Alliance] sooner, too. . . . I also have the feeling, I realize, as far as something as drastic, as illegal, as war tax resistance—I don't feel that it's for everyone. (. . . [E]ven though the more people that do it, the better . . . it's my feeling.) But I think about my sister and her husband and I don't want them to do it. I don't want their lives to get screwed up. They think a lot like I do about these issues and I feel like I want to do it for them so they can just live their life, you know. They have a little baby and I want them just to get to live. Not that, you know, I'm any kind of martyr or hero, that way, it's just a feeling of love for them and, you know, I'd like somebody else to do it for me so I could just live my life, you know?

Greg

Some years ago, another less active member, Greg, and his wife, Ann, were also active members of the Alliance. Around 1982, however, they moved out of the immediate Alliance area to a neighboring community farther from the city. More than ten miles from some of their Alliance friends, their business and personal responsibilities (they had two small children) also were expanding, and they were forced to cut back somewhat on their participation in the Alliance. Greg started an informal branch of the Alliance in their new community and he and Ann still attended Alliance events quite often, although they came to be only rarely seen at regular Alliance meetings.

Raised in northern Ohio, Greg got his law degree at George Washington University, and then he and his wife moved to California. Settling in the city near the Alliance area, he and his wife both worked (Ann as a registered nurse) but were able to get out of the city on weekends and frequently found themselves visiting the Alliance area. After working for two or three years, Ann went back to school to extend her nursing skills and become a nurse practitioner. Soon after her return to school, Greg also left regular legal work and began seriously to pursue a long-standing interest in art at a well-known school in the city. As they both neared the end of their respective training programs, Ann began to wish they could move out of the city. Although Greg was somewhat less enthusiastic about a move to the country than his wife, he was willing to move to the Alliance area he and his wife had visited on weekends, and after a six-month search for a house, this was the move they made.

Work was not as easy to locate as it had been in the city. Ann, in fact, couldn't find regular work for about fifteen months, a period during which she took several odd jobs outside her field. Gradually, however, by substantially broadening the range of her activities, she began to have better luck. She became involved in home health care, ran the local free clinic for a while, did some teaching, and worked some for a major "self-care" journal. Greg, thinking about getting into the real estate business, joined three other partners in the speculative construction of a house in the area, taking principal responsibility for the paperwork management of the project, although he also did some of the actual construction. When he and his wife bought the small farm they now live on, Greg also did much of the major renovation work on their own small home and on another structure that they later rented out as a separate residence. Unencouraged by the housing market in the area, Greg at least temporarily dropped the idea of going into real estate. He instead became a partner in a small local business that paints designs on clothing. This activity seemed to provide an outlet for both his artistic and his business/legal skills. The business, which was one year old and faced an uncertain future when he joined it, grew rapidly after he joined it and by 1983 appeared to be thriving financially and otherwise.

Greg's energetic, well organized, and incisive contributions to the business of the Alliance were clearly helpful and widely appreciated at group meetings in the early informal part of the observation period of this study. After he was no longer able to attend meetings, however, reference was often made, in describing his activities, to a wall poster he put together that located and described radioactive materials activities in the state of California. Greg's initial interest in this topic was stimulated by extensive discussion some years earlier of the possibility of legally designating certain areas as "nuclear free zones." As he became interested in this idea, Greg had first been surprised to find that there was no easily accessible collection of data on the full range of nuclear activities—i.e., on nuclear power production, on placement, research, and production of nuclear weapons, on radioactive waste storage, and on other radioactive materials sites, to say nothing of materials transportation routes. As he began to collect information of this sort himself from application-specific sources, his second surprise was at the ubiquitousness of these sites when all major applications are considered. It quickly became apparent that the designation of "nuclear free zones" would in most cases of interest require the discontinuation or removal of some existing radioactive materials activity. As a contribution to the Alliance's goal of educating the community, Greg put together a color poster that included a keyed map of the state with symbols at each nuclear materials location; the poster was printed and was still available at the Alliance office at the conclusion of this study.

Stylistically, Greg's speech was rapid, definite, and easily heard. He was clearly not inclined to waste anyone's time. His occasional indulgence in a humorous remark was, perhaps for this reason, often unusually successful and might be considered another of his contributions to the Alliance as a group effort. At a "Peace Charette" cosponsored by the Alliance that attracted forty-nine people from the community (many of whom were not closely familiar with the Alliance or its members), Greg achieved one such success. Ronald Reagan was entering the second half of his first term as president and sentiments in the community were running very high (as indicated, for example, by the number of people attending the "Charette") against his placement of cruise missiles in Europe and other aspects of his arms policies. At a particularly serious point in the Charette, separate working groups had joined to present their thinking to the whole group on questions such as: "What changes or events in [the community] could contribute to a 'World Beyond War' in the next year?" Greg broke a silent pause with an apparently serious proposal. He suggested that the community should petition to have itself designated by the federal government as a site for one of the new MX missiles. Greg had spoken to the group on several previous occasions and, taking him quite seriously, the group was silent for a moment as they seemed to try to discern how they might have misheard or misunderstood his suggestion. After a well-timed pause, Greg explained that having their very own MX would bring at least two benefits to the community: (1) in the event of an actual nuclear war, the community would surely be targeted and no one would have to hang around for the aftermath, and (2) with their own MX, members of the community would no longer have to travel the substantial distances to existing weapons facilities in order to participate in protest demonstrations. As a result of his style of delivery, Greg was well into this justification for his proposal before people began to understand that he was joking. The subsequent release of tension in the group was welcomed and the joke widely appreciated.

Peter

Another less active member, Peter, was one of the strongest advocates for renewable energy systems in the Alliance and one of the group's most articulate members. His own efforts to start a solar contracting business locally, however, had met with little success. When his first solar venture, intended to specialize in sun space (solar greenhouse) additions, domestic hot water, and other solar and conservation applications, began to have money problems, Peter borrowed several thousand dollars, mostly from his Alliance friends. As it then became clear that his business really was not sound, Peter's slowness in repaying his friend's loans became a source of some strain in his relations with the group. This strain and Peter's inability to stimulate the level of support for renewable energy systems among members of the Alliance that he felt would be appropriate, contributed to a sharp reduction in Peter's level of participation in the group. (Peter's unsuccessful effort to get the group's explicit designation of renewable energy systems as one of its primary areas of interest in an official statement of objectives was alluded to briefly in chapter 1. See also note 15 to chapter 1.) Once a regular participant in meetings and events, Peter came to be only rarely present at Alliance functions.

Still in his early twenties, Peter was one of the youngest participants in the Alliance. Although he left both school and home at a very early age (he later obtained a high school equivalency degree), he was still haunted by what he regarded as the bad "models" provided by the adults in his early life.

> I come from . . . real uptight, violent models, you know. . . . as bad as anything on television. And I'm still pretty reactive to that. I mean I can still get touched very easily in the wrong way . . . or touched in a way that I react as a thirteen-year-old kid. Which plays hell in [personal] relationships . . . if you're not all there for people. And so that takes . . . a lot of focus [and] quiet, you know, breathing and meditation, and just realizing that I am more than the sum of my experience.

Peter's father retired from military service not long after Peter was born and Peter's father and mother (now divorced) seem to have never had a good relationship. Without going into the details of any of this, Peter reported that he essentially "adopted a set of parents down the street" when he was ten or twelve years old. He briefly described his later activities en route to the Alliance community as follows:

> Oh, I made T-shirts and I sold drugs and I wheeled and dealed, and then I got tired of it and I came to [the city near the Alliance area] and I shot pool and I printed T-shirts and I sold drugs, and I wheeled and dealed. I saved up money and I bought a truck and tools and I ran [i.e., used the] truck and tools doing yard work and stuff. And then I started building houses [working in home construction] and I moved to [the Alliance community] to start building a house.

In 1983 and 1984, Peter was supporting himself through a wide range of construction work, almost all of it of a short-term nature. His age and experience were difficult to judge from his appearance (he is tall and wiry) or his behavior (which often made him seem much older than he was), and he was fairly successful in finding short-term jobs at as much as fifteen or twenty dollars an hour (e.g., doing electrical work over the Christmas holiday season). The flexibility of this kind of employment allowed Peter time to work on development of his own business, to attend conferences and study new solar equipment options, and occasionally to fast (a process whose "healing" effects Peter greatly valued) or just take a break.

Peter had accumulated an extensive collection of books and information relating to renewable energy systems that ran the full spectrum from works of social or national scope, such as Amory Lovins's *Brittle Power*, to detailed dealer's specifications for specific hardware systems. These materials took up a substantial amount of the space he shared in various locations (generally one or two small rooms) with his girlfriend, but they contained the kinds of information Peter sought and absorbed almost obsessively. He was also a regular participant in solar-related conferences at the local, state, and national level, generally avoiding registration fees by assisting with video equipment (e.g., at a passive solar conference in Colorado), operating slide projectors, or recording conference sessions and making tape copies available for sale.

Peter had learned a great deal in this way and, considering his age, experience, and lack of capital, he had made some surprisingly useful connections. When his own first attempt at a solar business failed, for example, Peter took the sun space designs he had found and had hoped to develop himself to an established partnership (about fifty miles from the Alliance community) with many years of experience in the commercial greenhouse construction business. He worked for this company periodically over a period of many months as they developed a prefabricated component approach to the design and began marketing it as a sun space addition. Peter had also located and directed them to a number of specialized products uniquely suited to the design (particularly caulking and weatherproofing materials) and was fairly well paid while he was actually working on the first few installations and other aspects of developing this new line of business. (He had no equity interest in the project, however, and, with limited interest simply in installing subspaces, he was not likely to benefit further from these efforts in the future.)

More recently, Peter began trying to launch a solar hardware distribution business. Hoping to get started by offering a very limited selection of devices, he collected catalogs and information on all of the operations of this sort (i.e., general solar equipment distributors—there were about six of them) and was trying to select new products that both offered good potential based on market needs and appeared to fill gaps in the product lines available from the other distributors. He was most pleased in this regard when he encountered representatives of a highly respected German firm at a conference and managed to talk them into giving him a try at an exclusive distributorship for the company's new flash (instantaneous) water heater. Believing that this type of heater could become a common component as a backup for solar water-heating systems and that the established name and good construction of this particular gas heater could make it attractive in this country, Peter was anxious to get this second business attempt launched. He had selected about five initial product offerings that he believed to have unique potential in the solar market (including the flash heater and one of the specialized glazing sealers he had used in the sun space work) and had begun talking with potential investors and business partners.

At times one might be tempted to write Peter off as a hopeless dreamer, and this was clearly a source of some of his difficulty in relating to other members of the Alliance. While he was very articulate, he was from a practical standpoint severely limited by his lack of business experience, his weak written communication skills, and the difficulty he sometimes had in working with other people. In addition, while he kept trying to start a business, it was never entirely clear that a business was really what he wanted.

> But it's actually . . . where my [first] business fell apart—that's when my vision of . . . what I wanted to do really started crystallizing. . . . Because I could see on a bigger scale, not just bogged down in the day-to-day bill paying and, you know, reputation salvaging. I could see a bigger scale than the little game that I had been playing for money. It's like am I doing this for money or am I doing this . . . to build models of a world that I'd like to live in, you know?

Peter's surprising fund of detailed practical knowledge and skills notwithstanding, he did occasionally drift off into an apparent never-never land of "community cable TV," home computer networking, and other apparently attractive images.

Established business interests would surely balk at Peter's discussion of his experience with psychedelic drugs[1]—experience he felt had helped him achieve a better balance in life—though of course he did not open such discussions in a business setting. For some years Peter had also been an avid follower of the musical group known as the Grateful Dead and he linked Dead concerts intellectually with his drug experiences. Both provided what he referred to as "real experiences"—experiences that were unexamined, direct, not filtered in the normal sense by the conscious mind. To the author's characterization of Dead concerts (I attended one) as a form of "pagan unity ritual," Peter reacted that Dead concerts were the closest thing to church he had ever encountered; noting that he was once an altar boy in the Catholic Church, he said: "Altar boy in the Catholic Church—*that's* pagan." For Peter, "Western man is very much caught up in intellect—maybe over the edge." If one side of the brain dominates the other, he once said, a person ends up in an "institution" (either a mental institution or a corporate institution, depending upon which side has become dominant). As he saw it, the Dead nourished one side of the brain while other normal activities in Western society nourished the other side.

It is not easy to imagine a successful marriage between this kind of thinking and investors in a small, new business enterprise. Yet Peter's intelligence and determination at both the practical and the theoretical level are undeniable. Some of his handicaps may even, in some ways, have been of use to him.

> I guess a lot of people [have] tried to teach me lessons, and I'm, I guess, pretty stubborn when it comes to learning lessons. I like to learn my own lessons. And I don't like people who feel it's their job to teach me a lesson.

While it would be difficult to predict the future for any member of the Alliance discussed in these pages, Peter's future could in many ways be the most elusive.

In Peter's view, much in community life had come to be "ass backward." He further believed, however, that the need for redesign was widely felt and that there would really be little trouble supporting a "planner" to coordinate redesign efforts in many communities right away if it were not for the simple fact that people had not become accustomed to paying for such services in this way. (In order for the traditional research and development approach involving the federal government, major corporations, and universities to be augmented by more local participation, the principal obstacle was simply that the decentralized approach would have to become "popularized" before people would feel comfortable paying for it.) Seeking a world in which it would again be possible to operate "in line, on purpose," Peter saw decentralized renewable energy systems as an important component of any design in which

> you enable people to learn new things and . . . take part in building their own pictures, their own city and town, without feeling that Big Brother is coming in and doing it for them.

3

Interpretation and Significance

CHAPTERS 1 AND 2 PRESENT DESCRIPTIONS OF THE ALLIANCE AS A group enterprise and of Alliance members as individuals. This chapter begins with a tentative practical interpretation of the study. This practical interpretation is then placed in several theoretical contexts with primary emphasis on Alliance activity as part of a value-oriented movement as described in the theory of collective behavior. Finally, the remainder of the chapter gives consideration to some of the implications of the failure in traditional energy policy settings to grasp the messages and hear the muted voices of people such as those who have been active in the Alliance.

INTERPRETATION

The Alliance and its members have exhibited a consistent pattern of preference for community and environmental harmony over more traditional personal, career, and short-term material interests. They have striven to protect these values (community and environment) from what they have regarded as the threats of offshore oil drilling, nuclear power development, and nuclear war or weapons proliferation; and they have tried to enhance these values directly through their activities in the community and through their efforts to develop more harmonious renewable energy and conservation alternatives. These values, and many of the beliefs and norms associated with them, appear for the most part to have been adopted and firmly internalized[1] by individuals before they moved to the Alliance community. The depth of conviction, the similarity and specific content of individually held values, and the sense of relative social isolation[2] these values have implied have all contributed to the success of an otherwise surprising consensus approach to decision making in the Alliance as a group.

The Alliance can perhaps best be understood, then, in terms of its members' affirmation of patterns of interpersonal and man/nature relations that depart markedly from those structured into traditional social roles and institutional processes and into the expressions of both that are embodied in the conventional design and arrangement of technology. This affirmation encompasses both a rejection of traditional patterns and an attempt to develop and embrace new alternatives.

With respect to human relationships, members of the Alliance have consistently sought to replace hierarchical, corporate, and market-oriented interpersonal relationships[3] with broader human relationships in which they are more able to share interests and express concern on a sustained basis. This effort is evident along with its technological component (and in spite of its immediate economic penalties) both in unpaid Alliance-related activities and in the unique character of paid work, such as Brad's appliance repair or Art's energy conservation business, within the community, both of which have depended primarily on personal and word-of-mouth contacts. Cooperative collegial activity has been valued above hierarchically organized endeavor, as evidenced by individual reports of, and reactions to, experience with major corporations and public institutions, and by the manner in which more active members of the Alliance have interacted with less active members in developing group projects. In these and other respects, Alliance members have frequently found paid work to be substantially less useful and less rewarding in human terms than work that appears not to be valued at all in the marketplace.[4] As one target of their activities, the deployment of cruise missiles in Europe and the general escalation of the arms race early in Ronald Reagan's presidency perhaps epitomized the kind of embodiment in technology of power-oriented, humanly remote relationships that members of the Alliance have found least attractive.

With respect to man/nature relations, members of the Alliance have consistently sought to express and effect a less aggressively rapacious relationship with the natural world both through their personal lifestyles and in their efforts to influence the choices made institutionally (e.g., among energy technologies) in their names. Although not previously described in all its detail, the deliberate selection of an attractive natural area in which to live and work has been accompanied in many members by a relatively heightened awareness of local topography, weather, microclimatic and other natural conditions. Unusually low personal consumption of energy and other resources is also implicit in the low incomes and simple living conditions of the group's members. Local, regional, and national environmental concerns generally provide good indicators of areas of concentration both for the group and for its individual members. To the degree that society's choice of technologies, and perhaps energy technologies in particular, lies at the very core of our practical expression of a relationship with nature, environmental concerns also lie at the heart of Alliance members' opposition to nuclear power, offshore oil drilling, and other classical practices in energy production and use. Similarly, part of the attraction of conservation and renewable energy alternatives appears to lie in the less disruptive or adversarial expression these alternatives appear to allow.

Alliance members' rejection of traditionally structured interpersonal and man/nature relationships has been strongly complemented in the substantive objectives to which it gives rise, by the endorsement of nonviolence and consensus decision making with respect to process and procedure.

THEORETICAL PERSPECTIVES AND COLLECTIVE BEHAVIOR

It is difficult to discern any relationship between the conventional policy-analytic approaches that traditionally frame our choice of energy technologies[5] and the behavior reported in this study. What might, for many citizens and government officials, be a strictly "rational" selection of least-cost energy production systems would for many members of the Alliance be a strictly short-sighted sacrifice of central values in the name of immediate convenience. Given Alliance members' deviations from aggregated expressions of market preferences, even the more fundamental bodies of economic and other theory underlying conventional policy-analytic approaches seem to shed little if any light on the behavior observed in the course of this study.

Other bodies of theory, however, including individual and social psychology, the study of group dynamics, studies of power and control and how it may be distributed locally, and so on, seem likely to be more helpful in interpreting Alliance behavior and its policy significance. Steven Lukes's profoundly insightful little book, *Power: A Radical View*,[6] for example, suggests that we might think of Alliance activities as evidence of a "third dimensional" exercise of power. Years of apparent acquiescence in the nonparticipatory technology design and development processes that brought us nuclear rather than renewable energy supplies and relatively wasteful rather than highly efficient patterns of energy use, for example, need not be taken, in Lukes' theoretical framework, as expressions of either preference or "real" interests. Drawing our attention to control over the political agenda and to the possibility of "nondecision making" in which decisions are effectively made simply by not bringing an issue up for consideration, Lukes might suggest that present energy technologies could have emerged and gained dominance over other alternatives primarily because those alternatives were never officially recognized or developed for popular consideration. Alliance activities themselves might be taken in Lukes's framework as clear evidence of long-standing latent conflict[7] over the design of energy systems that has surfaced only as authoritative positions have been opened to scrutiny by the unanticipated dislocations of the energy crisis. Traditional "pluralist" political theorists, concerned as they are only with influential groups in the political process and with the outcomes of overt conflict, will likely see nothing of interest in Alliance activities. Lukes's more probing analysis, on the other hand, begins to raise precisely the kinds of concerns about fundamentally undemocratic exercise of power in the shaping of technology that are central to this book.

For our primary interpretive framework, though, I want to focus on the theory of "collective behavior" or, more narrowly, the theory of social movements. This body of theory will bring some of Lukes's perspective to bear indirectly but also allow us to put the value-oriented movement that appears to encompass Alliance efforts in a general social and political context. I begin with a brief overview of the theory itself then proceed to its application to the case of the Alliance.

The central elements of the theory of social movements appear with some agreement[8] to be contained in three major works, written in the early '60s and '70s by Smelser,[9] Gurr,[10] and Turner and Killian,[11] respectively. The field has over time turned toward what is known as the "resource mobilization perspective" as represented in the works of McCarthy and Zald,[12] Barkan,[13] and many others in more recent years. This modern branch in the theory of social movements, however, appears to focus less on the content of social movements and the beliefs that may be incorporated in such movements and more on the tactical and structural determinants of their success or failure; while the latter issues are of obvious interest among movement organizers, politicians, and others, they are not central to the more content-oriented focus of this study. The theoretical perspective to be briefly reviewed then applied here is that of the earlier works on social movements, and among these reliance will be placed primarily on Smelser's early book, *Theory of Collective Behavior*.[14]

Neil Smelser defines collective behavior as "mobilization on the basis of a belief which redefines social action."[15] He distinguishes six determinants affecting both the occurrence and the specific form (e.g., panic, fad, social reform movement, etc.) of any given instance of collective behavior:

structural conduciveness
structural strain
growth and spread of a generalized belief
precipitating factors
mobilization of participants for action
the operation of social control

Arguing that these determinants need not become effective in any rigid temporal sequence, Smelser suggests that instances of collective behavior are progressively shaped as each of these determinants is brought definably to bear.[16] He further proposes (in accord with Talcott Parsons and others) that social action be considered in four components:

values
norms
mobilization of motivation into organized action (i.e., "the *form of organization* of human action")

situational facilities (i.e., "the means and obstacles which facilitate or hinder the attainment of concrete goals in the role or
organizational context")

These components of social action are, in Smelser's view, hierarchically organized in the sense that norms depend upon values,
mobilization into organized action must be derived from its associated norms and values, and so on. Without trying to reproduce
it in any more detail here, this framework provides the basis for
Smelser's description and analysis of a wide range of collective
behavior, including what he refers to as the "value-oriented
movement."

A "value-oriented movement," again in Smelser's formulation,
is defined to be

> a collective attempt to restore, protect, modify, or create values in the
> name of a generalized [value-oriented] belief. Such a belief necessar-
> ily involves a redefinition of norms, a reorganization of the motiva-
> tion of individuals, and a redefinition of situational facilities.[17]

With more specific reference to value-oriented beliefs, Smelser
states:

> A value-oriented belief envisions a modification of those conceptions
> concerning "nature, man's place in it, man's relation to man, and the
> desirable and nondesirable as they may relate to man-environment
> and inter-human relations." This kind of belief involves a basic re-
> constitution of self and society.[18]

This characterization of value-oriented beliefs closely matches
the basic interpretation of the Alliance offered in the preceding
section and firmly places the group in theoretical terms as a local
manifestation of broader value-oriented movements in American
society.

IMMEDIATE POLITICAL SIGNIFICANCE

Having placed the Alliance in the framework of the theory of
collective behavior, one aspect of the group's potential political

significance is readily apparent. If the Alliance constitutes a local manifestation of broader value-oriented movements—say the "environmental" movement, the "antinuclear" movement, the "voluntary simplicity" movement,[19] or more broadly among those sharing Perception C as outlined by SRI (see chapter 1)— what does the study presented in chapters 1 and 2 say about the potential for broad shifts in social preferences? Is it possible that perspectives not comprehended by conventional policy-analytic approaches could come to dominate decision making in the future, invalidating current analysis and planning in the process?

In essence, these questions boil down to asking whether the movement or movements represented by the Alliance will grow and ultimately succeed. As noted earlier, questions of this sort are not central to this study and would better be addressed in the context of more modern branches of the theory of collective behavior beginning with the "resource mobilization" perspective. A single community study might also seem a less useful approach to such questions than a broader social survey or other techniques aimed at a larger sample. Certain observations may, nevertheless, be of interest.

To begin with, it may be important not to underestimate the strength of Alliance members' convictions. From the broad coverage given to peace marches and other large demonstrations in years past, it may be tempting to conclude that large numbers of demonstrators are attracted more by a "party" atmosphere or the prospect of a free outdoor concert than by any serious belief that the causes they espouse will be furthered by their gathering. For members of the Alliance, however, music has provided an expressive outlet for almost obsessive commitments—an outlet that might in other groups and other movements be found in violence. The common practice of combining a visit with friends or a short vacation with travel to a demonstration similarly provides for an indispensable balancing and for moderation of an intensity of feeling that may only be fully understood among relatively powerless individuals whose values lie in fundamental conflict with the society in which they live. The individuals' stories recorded in this study amply demonstrate the potential for withdrawal as an alternative to capitulation in unacceptably standardized relations with other people and with the natural world. In this author's view, in fact, Alliance members' rejection of such socially standardized relations is probably best understood as being comparable to the body's physiological immune response in its rejection of foreign pathogens.

Further evidence of the strength of Alliance commitments may also be provided, in a certain sense, by the success of the consensus process itself. In ordinary circumstances, interpersonal and intergroup differences in priorities and interests arguably make consensus decision making impracticable; to hold action hostage to consensus is to take no action at all. Yet for the Alliance, consensus decision making does not perceptibly impede action. To the contrary, the group's central commitments are of such fundamental importance to its members that they appear to take priority in the minds of all participants over more ordinary personal differences, making action on central concerns uniformly preferred over delay on peripheral grounds of personal preference.

The persistence of Alliance and of individual member efforts may also be of significance, especially in the face of a lack of external rewards (profit, prestige, etc.) that would arguably be well within the reach of most members if they were to adopt other patterns of life. Organizationally, too, the group has functioned without the leverage of hierarchy or the inspiration of a single charismatic leader, again, on the basis of firmly internalized beliefs of a closely shared nature.

The success or failure of particular social movements supported by the Alliance will not simply be a function of positive achievements either. While it is true that individual institutions, public or private, do not bear the costs of rendering a single potential worker unemployable, or of failing more generally to enlist that worker's best efforts,[20] social institutions, collectively, ultimately do bear such costs. The mere fact that the capabilities of Alliance members have not been harnessed in support of conventional values undermines the status quo as it is challenged by the alternatives supported by the Alliance.

Finally, there is specific evidence in this study of a pattern suggested in A. F. C. Wallace's classic anthropological treatment[21] of "revitalization movements." Wallace defines a revitalization movement as "a deliberate, organized, conscious effort by members of a society to construct a more satisfying culture." Revitalization movements involve, as a central feature, the reformulation of mental images of society and of culture—i.e., of the "mazeway" employed by members of society as a road map to living within that society.

> The structure of the revitalization process, in cases where the full course is run, consists of five somewhat overlapping stages: 1. Steady State; 2. Period of Individual Stress [due to failures of the old "mazeway"]; 3. Period of Cultural Distortion; 4. Period of Revitalization (in which occur the functions of mazeway reformulation, communication, organization, adaptation, cultural transformation, and routinization), and finally, 5. New Steady State.[22]

As a specific example, the individual histories of Alliance members strongly suggest a pattern of similar individual responses to stress (i.e., to apparent failures of conventional "mazeways") followed by early communication (facilitated by moves to California) and mazeway reformulation efforts characteristic of Wallace's stage four.

While there is, then, specific evidence in this study to suggest that the movements supported by the Alliance might prevail, there are also indications to the contrary. Particularly, the group clearly has not yet managed to enlist support in the form of major social resource commitments. They remain barely able, in fact, to continue to scratch out a living and they continue to depend, in their movement efforts, on earnings from other activities. In addition, they have as yet been unable to define in full detail alternative patterns of life consistent with their values that could be implemented in a full-scale revitalization of society. While this second failure may stem in some measure from the first, it remains difficult to see how many of the values underlying social movements supported by the Alliance are to be realized in practice.*

This study alone does not provide any determining indication of whether the movements supported by the Alliance will succeed. In this sense it cannot be used to argue that traditional policy-analytic frameworks are mistaken and *must* be reformulated. The fact that these movements and groups such as the Alliance have not yet prevailed is, however, not simply a function of the

---

*The home power movement presented in the second part of this book appears to carry the detailed definition of alternative patterns much further, greatly easing this difficulty in the years since Alliance activities were observed.

ideas and values they represent. The ultimate success or failure of these movements—indeed the degree to which they are able to enlist social resources in their development—is also a function of processes of social control. And these processes are themselves of substantial political significance.

Returning to Smelser's theoretical framework, other lessons of possible political significance can be derived from a study of the Alliance. As the subject of the next section, these lessons begin to raise concerns about traditional relationships between citizens and policymakers and about the degree to which energy technology decision making measures up to democratic ideals.

PUBLIC POLICY, SOCIAL CONTROL, AND DEMOCRACY

Because the questions they raise are so basic and all-pervasive, value-oriented movements can be among the most disruptive forms of collective behavior. Depending upon the disposition of Smelser's six determinants, on the other hand, any given value-oriented movement may disappear virtually without a trace. Wherever the effects of a value-oriented movement are felt, however, "normal" social procedures are, according to Smelser, necessarily thrown into question; since values underlie norms, "normal" procedures, almost as a matter of definition, fail to apply. This aspect of value-oriented movements poses particular problems in the area of one particular determinant in Smelser's list, "social control," and these problems, in turn, are of particular interest with respect to the political implications of this study.

According to Smelser, the "containment" of a social movement typically "involves the selective closing of certain behavioral alternatives and the selective opening of others." In more detail,

> [containment] involves four kinds of behavior on the part of authorities:
> (1) Ruling out uninstitutionalized expression of hostility [ultimately through the actual or threatened use of force].
> (2) Ruling out direct challenges to legitimacy. This involves drawing a definite circle around those governmental activities which are constitutionally inviolable. . . .
> (3) Opening channels for peaceful agitation for normative change

and permitting a patient and thorough hearing for the aggrieved groups. [A practice referred to as "flexibility."]

(4) Attempting to reduce the sources of strain that initiated the value-oriented movement. [A practice referred to as "responsiveness."]

Referring to the first and second types of behavior together as "political effectiveness," Smelser notes at least three possible combinations of these four behaviors and their associated effects. First, if authorities are "flexible" and "responsive," value-oriented movements tend to: (a) disappear, (b) change into some less threatening kind of movement (e.g., a norm-oriented movement), or (c) "assume a value-oriented form which is containable within the system (e.g., an institutionalized cult, sect, or denomination)." A second combination, permanent political effectiveness, unresponsiveness, and inflexibility (i.e., permanent repression), "tends to drive the movement underground and then into passivity." Finally, as a third possible combination, "a period of repression followed by a weakening of [political] effectiveness . . . tends to drive the movement underground, or at least into an extreme value-oriented position, and then permits it to rise as a full scale, and frequently [violent] value-oriented revolutionary movement."[23]

Beyond noting that the Alliance itself does show some of the characteristics that might be associated with the term "underground," no attempt is made, here, to determine which of Smelser's three cases of social control best describes the situation at hand. If one accepts Smelser's theoretical formulation, however, and agrees at the same time that the Alliance is a part of a value-oriented movement, then the question of where that movement now lies with respect to Smelser's three social control alternatives clearly becomes an important one for policy-making purposes. If, for example, we live today under circumstances of political effectiveness and limited flexibility and responsiveness in the area of energy policy, what kind of impacts might we expect from runaway global warming or other environmental crises, energy supply disruptions, or other threats to political effectiveness in the future, and how might these impacts be affected by an increase or decrease in flexibility and responsiveness now?

Smelser's framework also suggests a very different image of

the world from the perspective of participants in the movement itself. In particular, political effectiveness, insofar as it may through the threat or use of force elicit the payment of taxes or other "constructive" contributions to previously established values, norms, etc., may be seen by movement adherents as a form of extortion. Ruling out challenges to established policies that are considered fundamental to the legitimacy of authority (i.e., "constitutionally inviolable") or restricting challenges, for example, to the confines of institutionalized forms of economic analysis, may be seen as direct forms of repression—despite any initial surprise on the part of movement adherents at this outcome of the functioning of an ostensibly free and democratic government.

The Alliance and its consensus process provide a reminder that society itself, and particularly modern technological society, relies for its very continued existence upon a basic social consensus. While, at this writing, the looming threat of terrorists in the Middle East, in Europe, and elsewhere in the world remains an ominous cloud on the horizon, it is difficult to imagine conditions that could lead *any* of the members of the Alliance to such extremes of desperation. Questions may remain, however, as to whether the terms and responsibilities of society in matters of energy, economy, and the environment have been fully discharged by the rule of the majority under the social contract to which the Alliance and its members remain party. To reduce the matter to simplistic terms, are present outcomes consistent both with genuine majority preferences and with the protections for minority perspectives intended under our constitutional system of democracy?

CLASSICAL POLICY MAKING AND THE ALLIANCE

A wise man once pointed out in simple English the fact that our attention is always and inevitably selective. In energy, and in other matters, "selective attention," "wherein faced with the complexities and paradoxes of real problems . . . people selectively ignore [potentially] vital aspects and concentrate on only one or a few simple features, as if those were the whole," is unavoidable. Simple survival, it would seem, "requires both selective attention and inattention, or we would choke in a froth of detail."

We see what we focus on, and can hear a bird's song above the city noise. The mother, oblivious to danger, rescues her child from the burning house; the soldier rushes to meet the enemy, the martyr to meet his god. Love is blind and memory selective, fortunately.[24]

Selective attention can, of course, be either conscious and, as a result, relatively easily altered, or unconscious and hence less easy even to recognize or describe. In the latter category, failures of imagination and the constraints implicit in the conditioning effects of individual and collective experience may figure prominently.

It would be surprising if classical policy-making frameworks were any more immune to the operation of selective attention than any other well-defined perspective, whether of a group or an isolated individual. Yet any process for shaping or choosing among possible technologies for energy production and use that fails to grapple with the prospect of the kinds of value differences suggested by the activities of groups such as the Alliance runs a dual risk. On the one hand, its analyses and decision making may fail to address the practical realities of the future. And on the other, its fundamental legitimacy may be thrown into question among those who do not share its analytical assumptions. If these risks are to be reduced or minimized, much more careful attention will need to be devoted to gaining a greater appreciation both for what the world may look like from other perspectives and for the ways in which present practices may seem at variance with the most fundamental tenets of democracy.

In closing, it may be that the members of the Alliance are best described in the words of one of Bill's favorite songs:

> We are the gentle angry people
> And we are singing,
> Singing for our lives. . . .

To their credit, it is undeniably a gentle *strength* that refuses to yield to the pressures of a mass age.

To be able to make one's own observations and to draw pertinent conclusions from them is where independent existence begins. To forbid oneself to make observations, and take only the observations of others in their stead, is relegating to nonuse one's own powers of reasoning, and the even more basic power of perception. . . . [I]f one gives up observing, reacting, and taking action, one gives up living one's own life.[25]

As members of the Alliance endeavor to construct a world they would "like to live in," it is difficult not to wonder both: (1) what that world might ultimately be like and, (2) whether their efforts are supported by the social resources to which, as members of society, they may have legitimate claim.

Part II: Home Power: A Model for Participatory Democracy in Technology Decision Making

4

The Home Power Movement*

THE SECOND STUDY TO BE PRESENTED IN THIS BOOK MAY INITIALLY seem entirely disjointed from the first. In describing the adoption in individual homes of small hydroelectric, photovoltaic (solar cell), and wind electric power systems, it is true that I will not refer directly to the Alliance or to any of is members. The image of home power activities I provide is also drawn primarily from the 1990s with data collection beginning around 1989, roughly four years after fieldwork for the Alliance study was concluded. And while Alliance activities of necessity gave a substantial emphasis to efforts to *obstruct* certain energy technology initiatives, the defining characteristic of home power activities is a devotion to the *implementation* of other (renewable) energy technologies.

I believe there is, however, a close similarity between the commitments of members of the Alliance and the commitments of those involved in the home power movement.** The home power movement may even be best understood as a close evolutionary descendant of the narrowly labeled "antinuclear" movement in which the Alliance played a small part. Unwilling to accept the human and man/nature relationships structured into traditional technological arrangements, those involved in both cases have struggled toward new expressions involving previously undeveloped design alternatives. As individuals across the country, often with experience closely paralleling that of the Alliance, have undertaken the development of alternative technological arrangements more consistent with their own value commitments, they

*Support for portions of this research has been provided by the Ethics and Values Studies Program of the National Science Foundation (grant number SBR–9511857).
**I continue to use the term "movement" in the sense of Smelser's "value-oriented movement" used in Part I.

have gradually found each other and combined their efforts in achieving the real successes yearned for by members of the Alliance and now evident in the home power movement.

Both studies also arise from the same underlying concerns with the social, political, and material implications of a prevailing pattern of impoverished public participation in the design and development of technology in the United States. Both are concerned with the necessity that we grant legitimacy and listen more carefully to the more muted voices at the periphery of traditional design practices if we wish to claim that the shaping of technology in our society is legitimate and consistent with the fundamental tenets of democracy.

To return for a moment to the themes outlined in the introduction, it has been noted that those concerned with the legitimacy of policy making in areas of science and technology may be more inclined to point to the "political construction"[1] than to the "social construction" or "social shaping"[2] of science and technology. To the degree that choice is involved in the shaping of science and technology—whether that choice is made actively or only by default—the shaping process itself is inherently political. In a democracy, we must attend carefully to how that choice is and should be made. This remains the core concern in both of the studies presented in this book.

Finding Our Bearings

Concerns about the legitimacy of the political construction of science and technology have only continued to be heightened since the completion of the Alliance study by maturing apprehensions of the ubiquitousness of the effects of choice in these areas in the shaping of ordinary lives.[3] Moreover, by the time formal work began on the home power study reported here, theoretical and practical questions about legitimacy had reached new levels with the sudden and unexpected end of the cold war and the substantial easing of its associated technically sophisticated military threats. Crisis-dominated rationalizations of science and technology policy dating back to World War II[4] were now quite suddenly open to question, at least in theory, as their foundations in military and national security interests were weakened or entirely

undermined. Although new crises such as that of "global competitiveness" showed signs of providing almost seamless transitions
to new justifications for old policy-making patterns biased in
favor of inputs from the "technically qualified," there was at
least in this brief transition period some additional stimulus for
reflection: who should be involved in deciding issues of science
and technology in a democracy and how should the decision-making process proceed?

Concerns of this sort also continued to be implicit in much popular discourse[5] and in NIMBY ("Not in My Backyard") resistance
not only to nuclear power but to many other officially sanctioned
technical initiatives from waste incinerators to genetically engi-
neered bovine growth hormone. They were evident as well in
scholarly interest in mechanisms for assuring or enhancing public participation in science and technology decision making[6] and
in European and other efforts with participatory technology assessment, "science shops," and "consensus conferences" involving lay participation.[7] Some recognition of the potential value of
citizen participation, even in traditional scientific terms, was beginning to emerge.[8] And a few scholars had initiated efforts to
identify more far-reaching models for that participation.[9]

Motivated by the legitimacy concerns revisited above, the second part of this book ultimately seeks to draw lessons from the
experience of the home power movement in the United States for
a more democratic approach to the design and development of
technology. As was the case with the Alliance, I will in part be
trying as honestly as I can to give voice to the otherwise muted
messages that participants in the home power movement themselves might wish to have conveyed.[10] As a part of this effort, I
will be describing some of the concrete developments of the movement that have carried its participants' alternative perspectives
well along the road toward actual alternative technological real-
izations. But my object will now extend beyond gaining a hearing
for any particular set of voices. My ultimate target in this reporting is to draw more general lessons for a *process* that would be
more open to muted voices of every sort.

These lessons will at the same time remain difficult to list or
explain without continuing reference to the details of home
power experience. Beginning with a broad brush, I will be describing significant individual and collective participatory re-

search efforts and a more generally reformulated politics of participation, as well as a loosening of classical corporate and economic relationships amounting in some cases to a partial withdrawal from the market economy. Altered patterns of access to information and a much closer relationship between the designers (experts) and users (laymen) of technology will also be prominent features both in our interpretation of home power activities and in modeling a more democratic process for the shaping of technology.

In more specific terms, I will begin by describing home power technologies and the movement itself in some detail, characterizing much of the technology development itself as the product of extensive participatory research efforts. Brief attention will then be given to how this unique movement could have emerged as it has from a background dominated by very different cultural norms and patterns of behavior. Unique levels of citizen activation and a heightened sense of problem ownership will be noted, along with distinctive patterns of participation in politics and the economy and in technology design and development itself. Finally, drawing from these practical interpretations and from theory, I will highlight certain features of a model that might be extracted from home power experience and speculate as to its possible utility and implications.

In theoretical terms, I will ultimately rely on Benjamin Barber's[11] notion of "strong democracy," and "more democratic" will generally mean "closer to Barber's strong democracy." Empirically, I will refer primarily to the NSF-sponsored field research reported here, including approximately fifteen weeks of participant observation and open-ended interviewing across the United States in 1995 and 1996,[12] although I will also be drawing from research on the home power movement I have conducted over a period of roughly ten years.[13]

HOME POWER PRACTICE AND TECHNOLOGY

The vast majority of home power installations in this country have come about through the close involvement of home owners in locating, designing, and installing or participating in the installation of their own systems. This pattern continues in the late

1990s and is, even now, necessitated to some degree by the lack of local retail outlets and installers and by the corresponding predominance of mail-order system sales. No precise count of home power homes in the U.S. is available, but estimates in the neighborhood of one hundred thousand as of 1993[14] appear to be supportable from sales histories and the experience of retail dealers over the previous ten to fifteen years. Of these, perhaps one thousand are grid connected,[15] and the remainder, most of which are some distance from utility lines, are not.

In terms of the technology itself, home power equipment has become quite reliable and very sophisticated in the degree to which it can be flexibly configured and even incrementally implemented for widely varying resource and end-use conditions. The vast majority of installed systems include some photovoltaic panels converting sunlight directly into electricity. A few systems rely entirely on microhydro or wind systems. And an increasing number are adding small hydro or wind systems to complement photovoltaic power supplies. These generation systems are typically accompanied by lead-acid (occasionally nickel-cadmium) battery storage and various battery charging and other interfacing controllers. Home power systems without any conventionally fueled backup generator are becoming increasingly common, and where such backup is included, it is generally called upon only a few times a year.

The earliest wind and PV systems first installed before 1980 served only twelve-volt direct current loads and required specially modified vacuum cleaners, kitchen mixers, and other appliances. Today, however, high-efficiency inverters producing pure sine wave alternating current even cleaner than utility power (i.e., freer of harmonic distortions) have become routinely available, frequently equipped with utility grid connection and many other options. New home power homes with standard 110-volt wiring have become quite common and, except in very small installations (e.g., vacation cabin), substantially outnumber twelve-volt systems among new installations. Several microhydro units optimized for different head (i.e., vertical drop) and flow rate conditions are now also routinely available, along with an assortment of small and very small wind machines. The smallest of the wind machines, rated at three hundred watts in a twenty-eight miles-per-hour wind, can even be utilized without a mounting tower. Sophisticated (pulse width modulated) charge controllers and pre-packaged metering and circuit breaker boxes are also available,[16] making systems more and more modular and owner-installation both safer and increasingly straightforward.

Particular system designs, generally arrived at through interactions between home owners and mail-order dealers, inevitably require that home owners at least specify in fine detail how they are and will be using electricity. Even that minority of home owners with no initial interest in the workings of their systems are pushed by dealers into such a specification, weeding out things such as resistance cooking and electric water or space heating along the way. An exaggerated commitment to energy efficiency and a unique awareness of energy use almost always accompany the adoption of home power systems as a way of making these substantially more expensive supply systems affordable. In some cases, specialized appliances such as gas-fired or superefficient electric refrigerators (the latter specifically manufactured for home power applications) accompany more familiar measures for reducing electricity consumption such as the use of high-efficiency fluorescent, in place of ordinary incandescent, lighting systems. (As of 1996, light emitting diodes or "LEDs" were also coming into use that are nearly ten times more efficient than fluorescents.)

From a business perspective, home power activity has grown to support on the order of ten multimillion dollar distributors and retail dealers[17] selling home power systems primarily by mail order. These businesses began in virtually every case from their owners' early efforts to develop home power systems for themselves and their own immediate neighbors. Their development has been aided, in many cases, by sales of home power equipment for use on boats, in recreation vehicles, and in communications and Third World development applications.

CONTEXT AND MOTIVATING CONCERNS

It must be emphasized, especially in view of the attention just focused on home power "technology," that what I am referring to as the home power movement did not begin historically with this technology. Nor is home power technology either the only or necessarily the most prominent technological or other innovation of those involved in the home power movement.

Historically, home power traces its origins to the counterculture of the 1960s and to the move "back to the land" that developed in that era, at least a decade before photovoltaic panels, for example, were generally available to the public.[18] In a story repeated from California to Arkansas and Wisconsin, young people essentially camping in teepees and old school buses in rural areas all across the country took the first step when they began rigging automobile batteries to play music or light a single twelve-volt bulb in their makeshift homes. They would recharge these batteries when they drove to town, sometimes shifting them in and out of the car on a regular basis, sometimes parking the car where they could jump-start it when the battery went down. The earliest improvements on this primitive and inconvenient electricity supply included efforts to refurbish Jacobs and other decades-old wind machines. Another early (but very maintenance intensive) approach was to jury-rig an old car alternator to a small garden tiller or lawn mower engine. A strikingly large proportion of home power pioneers and those who continue to play a primary role in home power businesses tell essentially the same automobile battery story,[19] and historically the home power movement is only secondarily an effect of the emergence of home power technology. Rather, the home power technologies we see today have emerged through a decades-long developmental effort in response to apparently widely felt needs and in support of different patterns of life.[20]

In the broadest terms, the home power movement appears to flow from its participants' desire to reshape their relationships with other people and with the natural world. Deeply felt environmental concerns, a reformulation of the relationships of community and of the workplace, and a shift in work content toward greater diversity in work activities all are strongly represented throughout the movement.[21] In keeping with the broad nature of movement concerns, technological and other innovations are by no means limited to the energy supply and end-use technologies of home power. From the earliest days, community-based and family "home schooling" of children, partly necessitated by remote locations, has been common. Cooperative road maintenance and a sharing of water supply, radiotelephone, and other responsibilities have also been common, especially where clusters or communities of home power homes have emerged. There remains today a very high proportion of owner participation in the construction of homes, as well as in the design and installation of home power systems. Owner-builders frequently make use of local stone, wood, and other materials, even cutting boards with their own or a neighbor's portable sawmill. In rural clusters, neighbors often work together on home construction.

As basic shelter and energy needs have been met, other innovations have also begun to mature. Extensive gardening as an avocation, a source of food, and a form of part-time self-employment continues to be widespread but is also increasingly successful and sophisticated, gradually developing toward locally tailored models of "permaculture" or "sustainable agriculture."[22] While the earliest communal living arrangements in rural areas across the country have essentially disappeared, closely cooperative communities of now well-established middle-class families remain. In new areas, innovative "land trust" and other institutional arrangements for ensuring more environmentally sensitive development patterns have begun to emerge and incorporate home power systems, small-scale agriculture, and other innovations for sustainable living.[23] A number of nonprofit and educational organizations have been formed, offering courses on owner-builder construction techniques, home power systems, and other topics. Several of these non-profits also have well-developed commitments to furthering the use of home power systems in Third World development.[24] And, again as basic needs have gradually been assured, the expression of broad interests in energy efficiency, electric vehicle design, and other issues has continued to develop. Extreme thermal efficiencies and advanced masonry woodstoves are now increasingly common in homes, for example, even in remote areas where fuel wood is easily and cheaply obtainable. And things such as electric vehicles, ranging from kit conversions and custom designs to lightweight electric mopeds[25] are encountered in significant disproportion to population.

From the earliest days, home power people have quite evidently worked much of the time from inchoate desires and conceptions of the world—from desires and conceptions of the world that conventional discourse and technology (seen as a form of language) have in fact afforded little *means* for expression. Specific explanations of what they are about are often framed in the widely understood terms of vigorous environmental commitments, and even in terms of the economic attractiveness of home power systems, although this argument depends entirely on the economically indefensible decision to locate in sometimes extremely remote areas.[26] Explicit ties are sometimes drawn, as well, to the slightly more revealing, but still rather vague language of the "appropriate technology" movement. Steve Troy, for example, traces the origins of his business, Jade Mountain, back to 1972 and a natural-foods store in California that began stocking homesteading supplies in response to customer demands.

> Jade Mountain began in 1979 and later narrowed the focus to Appropriate Technology products. Today our catalogs describe tools, ideas, and energies that fit E. F. Shumacher's code for problem solving in today's world: small, simple, inexpensive, and non-violent.[27]

There are also indications, however, of uncommon technical insights underlying, for example, environmental commitments. One small-scale retailer, for example, reports:

> My fundamental driving force is to decrease the amount of junk that people have to buy and consume and throw away. [And] that *includes* PV panels, inverters, batteries, wire. Ultimately, it's all garbage in the environment. . . . None of it lasts forever. And none of it is free as far as manufacturing and transporting it. It all gobbles up lots of resources. And it all dirties the planet. . . . [R]eally, my driving emotional force is I don't want to see more garbage on this planet. That's why I took interest in this field from the very beginning.[28]

There are occasional indications, as well, of a more sophisticated awareness of the extent of departures from mainstream culture. One small-scale dealer and installer in Kentucky, for example, explicitly recognized that home power systems are not "exciting and flashy" in the way we expect the latest new technology to be. Explicitly trying to move people away from this sort of false enticement, he expressed annoyance with occasional attempts he saw to make home power technologies appear to meet traditional expectations along this line.[29]

5

Home Power as Participatory Research

MUCH OF THE SUCCESS OF THE HOME POWER MOVEMENT IS THE product of a prolonged and extensive "participatory research"[1] effort. With its specific renewable energy elements formally chronicled since 1987 in the pages of *Home Power Magazine*, certain aspects of this effort are now increasingly supportable in the marketplace, but they have been driven overwhelmingly by personal commitments rather than traditionally recognizable interests. Seen as a product of participatory research, home power efforts continue to be advanced both by relatively isolated individual initiatives and by closely cooperative efforts, each of which is illustrated at some length below.

INDIVIDUAL INITIATIVE

Among relatively isolated participatory research efforts, one example would be the water pumping developments of Wes Woodford.* As is generally the case, the background for these efforts stretches back at least twenty years, beginning in the mid-1970s when Woodford and a few others started combing the countryside of the Great Plains scavenging old wind hardware from the '30s, '40s, and '50s. Learning mostly from the older people who had once used these systems, Woodford quickly became widely respected among the fifty or so small-scale dealers across the country who became involved in refurbishing old wind machines. In 1979, however, he shifted his attention to photovoltaic systems as PV panels first became generally available to the public at afford-

---

*I am again using pseudonyms preserving gender and referring to actual people I have interviewed at length, generally on several occasions over a period of several years.

able prices. (As he tells the story, this transition allowed him to sleep comfortably through a windy night for the first time in years as he began to worry less about somebody's wind machine coming to grief.)

Installing systems for himself and gradually establishing a small home power business, Woodford's interest in water pumping arose through direct interactions with his customers. Visiting customers very happy with their home power systems, he would often find them running their backup generators even in very good sunshine conditions. They would be using water in their gardens or elsewhere, it seems, and had to run the generator simply to pump water (a major energy user where wells are at all deep). Really good inverters did not begin to be available before 1986 and it was difficult to run standard well pumps without the generator as a 110-volt source. Ordinary well pumps also proved to be very inefficient, requiring the generator's large infusion of energy.

In talking with the pumping industry, Woodford was initially told that pumping without the generator was really impossible— and it was, in the sense that the kinds of pumps he sought were not available on the market. But traditional pumping professionals were not closely familiar with the old (e.g., wind-powered) pumping systems Woodford had seen, and he set out single-mindedly to develop a line of domestic water pumps that could be run on direct current and perform at much higher efficiencies than commercially available well pumps. Among the early systems he produced was a "pump jack," a reciprocating rod piston pump that looks like a small-scale version of a classic oil-well pump. Fitting these with direct current motors, he was able to deliver water with one-quarter of the energy used by a normal AC pump run from an inverter. PV-powered pump jacks are now widely used in home power homes and also in remote cattle-watering systems across the Great Plains. Among other innovations, he helped popularize the "slow pump," a positive displacement pump operating at varying flow rates directly from a set of PV panels—something he says he first learned was possible from one of his customers. In this system, a "linear current booster" is used to maintain necessary torques at low insolation (hence low PV power) rates. Water is pumped at a rate determined by the sunshine into a storage tank that buffers variations in actual domestic water use.

Woodford eventually hired a mechanical engineer and designed and patented a highly specialized deep-well pump, finally carrying his efficiency efforts beyond all market prospects. But he subsequently scaled back to a steady business specializing in pumping for home power systems. Monitoring and contributing to developments on an international scale, Woodford added an extremely efficient deep-well pump designed in Italy for Third World applications to his product line in 1996. He has returned to what amounts to a one-man business and appears to prefer to remain small, focusing his efforts in a way that obviates the need for additional employees.

COOPERATIVE EFFORTS

One of the most striking illustrations of cooperative participatory research in the home power movement has been the gathering of efforts in a home power community scattered in and around Amherst, Wisconsin, a town whose landscape is still dominated by wooded acreage and by the fields, lush green in summer and snow-covered in winter, of small family farms. The history of any of a dozen or so major figures might be traced as an introduction to efforts in this area, and a complete history would be deserving of book-length examination. For present purposes, three profiles of local residents and a brief account of the community's annual energy fair will have to suffice.

One resident, call him Roy, purchased an old farmstead in Amherst in 1971, just after finishing a degree in natural resources at the University of Wisconsin at Stevens Point (about fifteen miles from Amherst). He lived there communally with a group of other young people and became involved with a number of small-scale agricultural projects, including solar crop drying and the growing and grinding of organic grains. During the energy crisis of the 1970s, Roy began selling woodstoves from his home. He had taken an interest in an air-heating solar collector one of his professors had built back in 1971 and also went to work with a local solar heating business, eventually custom fabricating air collector systems, many of which were retrofitted to the exterior walls of existing homes. When the loss of tax credits for solar systems severely limited prospects for this activity, his partner in the business left, but he continued, paying his way increasingly from woodstove sales and an added line of home power products.

After twenty-five years, Roy still lived in the now refurbished farmhouse with his wife and two children, high school and college age as of 1996. (Only one other original member of the commune, who remained one of Roy's best friends, still lived nearby with his own family.) Over a period of ten years or so, Roy and his wife had continued to build the business to the point of employing about half a dozen people. They then sold it to Real Goods Trading Corp., a large, mostly mail-order, home power dealer based in California with additional retail stores there and in Eugene, Oregon. With Real Goods underwriting renovations to their store, a former lumberyard building, they continued to operate as one of Real Goods' three retail outlets and remained a primary resource for home power, energy efficiency, and hearth products (wood heating) equipment in the Midwest region. They also served as Real Goods' mail-order center for hearth products nationwide. Their home continued to be heated by a combination of wood and some of Roy's original air collectors, and was powered by a photovoltaic system. Except in the snow season, Roy also drove his own electric car charged by a large PV array at their house.

Clark, another Amherst resident, joined a separate commune several miles away after one year studying sculpture in college, arriving around the same time that Roy did. With some construction background (his parents had built their own home), he was soon involved in the construction of a number of homes in the area, including his own. With two partners and several other employees, he eventually incorporated what had been (and actually remained after incorporation) an innovatively informal construction business. Clark and his partners have specialized in home designs that combine passive solar features with hydronic slab heating supplied by active solar water-heating panels. These designs have gradually been refined, essentially through a series of experiments with his own home and those of his friends-neighbors.

In many respects Clark's own home has been one grand continuing experiment, beginning with thermal systems and continuing from around 1980 with home power systems. (His and several of his customer-neighbors' homes have never been grid-connected.) As a part of his home power efforts, he has also honed his own superefficient freezer configuration, adding insulation to a Sears chest-type freezer and removing its compressor and condenser coils, which he has augmented, some distance from the chest. (These heat-rejection components of a refrigerator or freezer are typically built into the unit in such a way that their heat warms the cooled space and reduces the unit's efficiency.) His thermal and electric energy systems are now fully functional, but experiments continue in the form of a new one-thousand-watt wind machine he installed around 1995, partly with the intention of reducing his use of a backup generator.

Carl also added a wood-fired masonry cookstove of his own construction to his kitchen around 1995. Masonry stoves had been among his most recent developmental interests and had been included in his company's construction projects as well as his own home. Relying on massive masonry structures to absorb and distribute the heat of a wood fire evenly over a period of many hours, these stoves allow homes even in northern Wisconsin's severe winter weather to be heated entirely with a single fire in the morning and another in the evening, much reducing the burden of keeping a traditional cast-iron stove full of wood throughout the day and into the night. Carl and his coworkers published a brief summary of their masonry stove investigations in *Home Power Magazine*,[2] sharing results, as contributors to the magazine regularly do, with home power folks all across the country.

Karen, a third local resident, moved to Amherst in 1979, after staying for a while with her sister and brother-in-law in Amherst. Falling in love with an eighty-acre parcel near her sister's home, she left her East Coast job and joined forces with Carl and her brother-in-law to build her home a quarter of a mile or so from the power line. She, Carl, and a third person in the area joined in the first PV purchases in the community in 1980. She, like Carl, added a one-thousand-watt wind machine to her PV power supplies around 1995 and was clearly pioneering this course with Carl in the Amherst community. In her case, she sold her backup generator at the same time that she installed her wind machine, hoping to operate entirely from the wind and PV systems. Carl reported in 1996 that "the jury is still out" in the community on whether wind systems like theirs would prove to be a good idea. And Karen, in this as in other stages of system development, was explicitly looked to by many in the area as one who was willing not only to try things out first but to stick with them as the inevitable bugs were worked out. In January 1996, a turnbuckle let go in one of the guy wires for her eighty-four-foot tilt-up wind tower, and the tower and wind machine fell to the ground, damaging at least the blade mounting plate and one of the blades. Roy was out early the next morning and, true to form, her question was essentially, "Well, what do we do next?"

When I interviewed her, Karen worked half-time as a medical technologist in Stevens Point and cut her own wood with neighbors for backup heating. On a personal note, she had recently turned down consideration for a supervisory position, not wanting her relationships with coworkers to be altered as she had seen happen in other cases. She also recognized that managers cannot make everyone happy and simply did not want to be the one forced to dish out the inevitable unhappiness. Notably, Karen was not very technically inclined but had developed a good schematic sense of how her systems worked. She strongly appreciated the links they provided her with the outside world and enjoyed keeping a log of sun and wind conditions and keeping mental track of the implied state of her energy supplies even though, as she herself noted, she had recently added sophisticated meters to her system that would allow her to check energy availability directly at any time. Karen was genuinely uncomfortable addressing groups of people, noting that this is difficult not only because one cannot monitor many people's reactions simultaneously, but because people in a group also seem to feel less responsible for showing their reactions than individuals ordinarily do. She had, nevertheless, risen to the occasion to assist with some of the group tours that regularly visited her home and even gathered her nerve to prepare and read formal testimony once at a public hearing.

In the immediate vicinity of Amherst there were around a dozen fully home-powered homes as of 1998, one of the most recent of which was formerly an Amish farmhouse that had never been connected to the grid. There are also a number of other energy-efficient and solar homes, and some of their owners plan for a shift at some point to home power as a source for electricity.

These are only indications, however, of the scale and duration of participatory research efforts in the area that come together most visibly, perhaps, in the annual renewable energy fair the community has sponsored every year since 1990. Now formally organized by the community's nonprofit Midwest Renewable Energy Association with an annual budget in excess of one hundred thousand dollars and three part- or full-time employees, this fair draws on the efforts of approximately one hundred volunteers and attracts equipment displays and workshop instructors from all across the country. Held at the local county fairground, mostly under circus tents, the three-day fair costs only fifteen dollars to attend and is about evenly divided between equipment displays and a series of ninety one-hour educational workshops on topics ranging from the selection of an inverter or installing your own microhydro-electric unit, to straw-bale construction, hydrogen production for energy storage, and energy curricula for secondary-school teachers. This fair, more than any other single event, amounts to a pooling of research results not only from the Amherst community but from across the country. It is also the site of active design interactions contributing directly to the emergence or continuing development of short-range, lightweight, two- and three-wheeled electric vehicles, a top-loading horizontal-axis washing machine (saving water and energy), and other technologies responsive to the interests of the home power movement.

The participatory research aspect of home power activities in the Amherst area has become increasingly explicit, in many cases coming to be consciously recognized. After considering the purchase of a Sunfrost refrigerator-freezer (the superefficient unit specially made in Arcata, California, for home power applications), for example, another resident, Peter, chose instead to get another nearby home power person to retrofit a standard unit for him. This involved removing and using the inner cabinet, adding very thick walls of insulation, building the unit into his kitchen cabinetry, and removing the compressor and condenser coils for reinstallation some distance away from the chilled spaces. This was still something of an experimental process even in the Amherst area and ended up costing as much in hardware and the local retrofitter's time as the rather expensive new Sunfrost would have. Remarking this as the unit was finally completed, both Peter and the fellow working with him were initially upset.

As Peter tells the story, however, Carl pointed out to him that he needed to consider what role he intended to play in the development process and recognize that if he wanted to participate in the early stages, he would have to expect this kind of experience. With this clarification, Peter and his wife, who remain grid-connected but supply the bulk of their energy from a PV system, accept with others in the community the added costs associated with what amounts to a participatory research and development process. (Professionally, Peter earns a living as a children's musical entertainer with an environmental education focus. He performs in public schools and other settings and has made several commercially available recordings.)

Consuming Avocation

It is perhaps worth emphasizing that the participant researchers who have brought us the home power movement and continue its development today have long worked in the lower middle-income brackets, often falling well below the poverty line and surviving only because of their remote location and the extremely low living expenses that often are an explicitly recognized and cultivated means to the development of alternative patterns of living. It is true that the movement has reached a point today at which a few of the early pioneers and a number of newly hired employees (e.g., at Real Goods, Sunelco [now a part of Photocomm], Jade Mountain, etc.) actually earn a reasonably secure living with health insurance and other traditional benefits. A number of the original pioneers have also managed over a period of fifteen or twenty years of learning, experimentation, and slowly accumulating assets, to get to the point at which their homes are reasonably finished and quite comfortable and attractive by traditional standards, although cash incomes may remain low even in these cases. The emergence of home power as a business bears very little relation, however, to classical business development models. In most cases, the businesses now in place came about accidentally: in the course of helping friends and neighbors set up their systems and get reduced prices by grouping their PV purchases, various remote home power developers essentially "found themselves" with businesses. Widespread surprise at the discovery of the viability of home power businesses was jokingly but well expressed by one pioneer still working part-time in a retail store in California. "Never would have 'thunk' it," he said.

Home power's participatory research efforts have been essentially unfunded and generally take the form of a consuming avocation, often weathering what would widely be regarded as extreme financial privation. Richard and Karen Perez, to take one example, have produced *Home Power Magazine* from their home, which they themselves have referred to as the "plywood palace," on an annual income that for many years did not exceed eight thousand dollars. They have recently added a tiny room to this same structure for the computer equipment they now use to prepare new issues on laser disk to send to their printer, but they continue to live essentially at the same level. Many small-scale home power dealers and installers continue to get by on comparable incomes in Kentucky, Arkansas, California, and other parts of the country.

Home Power Magazine is, itself, worthy of note as a pivotal link in participatory research efforts, spanning as it does, the geography and most of the history of the home power movement. Known and read almost universally among home power enthusiasts, it was initially free, printed on newsprint with costs covered by very modest advertising revenues. In the early years, one of the chief distribution modes was simply to leave copies of the magazine in laundromats in the knowledge that those who were struggling along trying to invent ways to live more independently ("on the land") generally did not have their own washers or dryers. Some of the earliest links among home power pioneers were established in this way and continue to flourish in the vigorous sharing of problems and solutions that occurs in both the articles and the letters published in *Home Power Magazine*. The continued existence of *Home Power Magazine*, as well as the devotion of its readers (who remain its chief source of articles) are just two indicators of the shared level of commitment to the participatory research process that has formed a foundation for the home power movement.

6
Origins and Explanations

It is DIFFICULT AT FIRST TO IMAGINE HOW A PARTICIPATORY RE-
search effort of this sort could have come about. As indicated earlier, however, elements of an explanation appear to flow from fundamental disjunctures between home power people and
mainstream technology and social institutions dating back to the
1960s.

In a Clearing

For one thing, home power has emerged in a unique and remarkable clearing that appears to have opened temporarily in
terrain ordinarily preempted by the dominant political and market economic institutions of our time. Until the mid-1990s, no
major market players apparently saw any significant prospect for
interest in home power systems, and the terrain, as a result, remained wide open to pioneers entering without organization, capital, specialized knowledge, or other traditionally recognizable
resources. Similarly, government long embraced the position that
PV and other renewable power systems would only become a rational choice if and when their cost per kilowatt-hour fell to a
level competitive with utility power. Wind research has emphasized large wind machines installed only in the most desirable
wind sites for utility and commercial "wind farm" applications.
Research and development in the area of photovoltaics has been
focused on such things as alternative cell types and materials and
on manufacturing process improvements designed to reduce
costs. The occasional government-sponsored PV home demonstration has included conventional, highly inefficient home appliances and covered essentially the whole roof of the house with the

sort of ten-kilowatt PV arrays that might be expected with very low cost panels. (Actual home power PV systems are typically one kilowatt or less and incorporate very vigorous efficiency improvements in electricity use.) Since home power homes have typically been built in rural areas without financing, even banks and building codes have played little or no role until quite recently.

Around 1995, these institutional "clearings" actually began disappearing rapidly as major players such as British Petroleum (BP Solar) and a number of electric utilities (Southern California Edison, Sacramento Municipal Utility District, Idaho Power, and Arizona Public Service, among others), began entering the home power market directly, sometimes joining with PV and other component manufacturers to sidestep the existing dealer infrastructure entirely. As the equipment has become more sophisticated and reliable, increasing numbers of mainstream home power homes have also begun to be built and the influences of bank financing, building codes, and Underwriters Laboratories approvals are beginning to be visible. There is even debate about possible certification of system installers. But, again, from, say, 1979 to 1994, the traditionally dominant institutional actors of government and the marketplace, from major corporations and regulatory agencies to banks and universities, were for practical purposes absent from the home power scene.

SHARED EXPERIENCE

At the same time, those who were to become the pioneers of the home power movement were in many cases finding themselves profoundly alienated from the mainstream of American life and technology. Among the dozen or so leading figures in the movement today, at least three became disaffected and left elite engineering schools before graduation, one after a dean labeled him a "communist" based on his early participation in civil rights and antiwar activities, informing him that he could not expect to get a regular job even if he graduated. Virtually all of these leaders are in their middle forties to early fifties, and virtually all were apparently vigorous opponents of the Vietnam War. At least one had made the decision and informed his parents that he would move to Canada rather than be drafted; a doctor's letter got him a deferment in a manner that he felt others from less privileged backgrounds unfairly missed. Another remains, with his wife, an active and committed Quaker and was deferred as a conscientious objector. Yet another was drafted and sought to refuse induction but got a medical deferment along with many others at his Oakland, California, induction center as, he believes, it worked to reduce the embarrassment of massive numbers of draft refusers. This last individual had been reading Trotsky and genuinely expected a full-fledged revolution in this country. He credits his Haight-Ashbury experience with rescuing him from this frame of mind and helping him to take a very different course out of the fray of the conflict. (Interestingly, at least three of the dozen or so leaders in home power today, including two of the most technically inclined, actually got their college degrees in philosophy, psychology, and anthropology, respectively.)

Many of those who have become involved in home power report that they had had substantial backpacking experience and were therefore aware that it was possible to live much more simply. Their experiences also appear to have heightened environmental commitments that were near their popular apogee in society at large as the baby-boomer pioneers in home power passed through their college years. Often aquiring relatively large pieces of land very inexpensively, they generally made their moves out into remote rural areas without possessions other than their tents and vehicles and had no long-range plans even as to how they would make a living. In many cases, log cabins, yurts, underground homes, and other innovative structures were eventually begun (these often remain to be seen near early pioneers' more fully developed homes today), also without specific plans regarding future electric power connections or other utilities or "improvements."[1] As one small-scale home power dealer now puts it, they had launched one grand participatory "lifestyle design"[2] project that continues, though much matured, even to this day.

Activation and "Problem Ownership"

It may be the combination of a unique "clearing," relatively free and open with respect to traditional institutional players,

and the often powerful disaffections or "alienation" of eventual
home power enthusiasts that begins to explain the early emergence of the home power movement. This unique combination
may also begin to explain the initial and continuing "activation"
and sense of "problem ownership" that undoubtedly remains the
most obvious distinction between participants in the home power
movement and the stereotypical "citizen" who now seems increasingly to be thought of in the more limited role of "consumer."

Spending time with them in Amherst or in other areas, one is
struck by the degree to which participants in the home power
movement are active in seemingly every issue area directly affecting their lives and by the degree to which they have become direct
participants not only in resolving but in defining and claiming energy, environmental, and other problems as their own. The sense
of collective empowerment associated with this activation, and
the effectiveness of efforts to act from a sense of problem ownership, are perhaps most obvious in a setting such as Amherst,
where enough people have come together in one place to provide a
critical mass for activities such as the Midwest Renewable Energy
Fair. The vigorousness and durability of their activation and
sense of problem ownership may be even more remarkable, however, among relatively isolated home power dealers and home
owners who, on a day-to-day basis, continue to weather financial
and other hardships largely on their own. In both cases, the sense
of recovered "agency," as it were, is palpably evident even where
it may seem most besieged.

The briefest exposure is sufficient to demonstrate that the
home power movement is not populated by the stereotypically ap-
athetic, nonvoting, and fashionably cynical citizens sometimes
denigrated by traditional policymakers as those same policymakers despair of any prospect for popular responsiveness in the face
of pressing public problems. These are not the kinds of citizens
who must be prodded and cajoled into getting someone simply to
blanket their water heaters for energy conservation or who abandon all thought of action as they await the arrival of suitable tax
inducements. Yet they also are not merely a collection of "tech-
nophiles" fascinated, in the fashion of modern computer hackers,
with new technical possibilities. John Schaeffer, an important
home power pioneer who has built Real Goods Trading Corp. into

one of the largest retail and mail-order home power businesses, was an anthropology major in college and was, himself, astounded when he first learned that he could purchase twelve-volt light bulbs at gas stations and hook them to automobile batteries to provide lighting in his own early shelter in the California hills. Many home power adopters have also learned what they know from scratch, in the course of putting in their own home power systems, and continue to operate from schematic understandings without real technical inclinations of their own.

Alienated from a world of pollution, war, and corporate enterprise, those who have produced the home power movement at first found little at hand in the way of alternative models or supporting technologies for another way of life. Essentially left to their own devices, they have embraced the only positive alternative open to them, becoming active and inventive themselves in what has amounted to an open-ended participatory research process. The disjunctures of the Vietnam era, though important to an understanding of the origins of the home power movement, almost never surface today except in response to direct questioning. They tend to be treated as a kind of ancient history, not insignificant, but now sustaining a more narrow resonance primarily in the remaining environmental and other values that continue to motivate a vigorously positive agenda.[3] In this context, technological alternatives have arisen not to compete as substitutes for other products in the marketplace (as, for example, natural gas or nuclear power might in the production of electricity) or even as innovations intended to initiate new markets (as with recombinant bovine growth hormone, "rBGH," or other products of genetic engineering) but as direct expressions of alternative values and as instruments essential to a more desirable pattern of life. They have arisen not in isolation, or as a response to corporate (profit), governmental (particular agency), academic (disciplinary), or other institutional drives, but as an expression of direct participation taking the form of a new fabric of technological means. As well-established middle-class members of their communities, no longer the least bit radical by most appearances, home power enthusiasts yet remain committed to a new pattern of life and to a level of direct participation in its expression and development that is, by contemporary standards, quite rare.

From all appearances, early feelings of alienation have been entirely displaced by the fascination of realizing the prospects of a new vision. As one home power person has put it, "I think that basically we have begun to believe that we really can [re]design our lives, you know?"[4]

7

Distinctive Patterns

A NUMBER OF DISTINCTIVE FEATURES OF THE HOME POWER MOVE-
ment may be worth highlighting as possible building blocks
toward a model for more democratic technology design and development. In keeping with the origins of the movement, for example, participation in politics and participation in the economy
remain distinctive in potentially significant ways. Expertise, also,
seems to be handled differently, at least in the sense that the gap
separating experts from laymen is much more narrow than is cus-
tomarily the case with science and technology design and development. There appears, moreover, to be a driving commitment to
education and a free flow of information that runs contrary to a
mainstream culture in which information may appear to be collected and controlled in such a way that its benefits become a
source of income for those who control it.[1] These distinctive features are developed in greater detail below.

PARTICIPATION IN POLITICS

Interestingly, the unique level of "activation" or "participation" characteristic of the home power movement does not generally extend far into traditional electoral politics, as might
ordinarily be expected. In a positive and prospective sense, it remains very local and focused on the building of homes and energy
systems, the schooling of children, and the other work of the community. On a larger scale, political action of a traditional nature
seems to appear only in response to direct challenges from the
outside. A proposed power line and the prospect of siting a nuclear waste dump, for example, both brought overwhelming (and
entirely effective) responses in the Amherst area. In other in-

stances, responses have been mounted to overgrazing of nearby federal lands and to other environmentally destructive local practices. But the grass-roots organizations that spring into action on such issues tend to disappear, at least in a superficial sense, when the threat is overcome. Actually, the politically savvy membership of groups such as those in the power line and nuclear waste dump cases quite consciously retains a latent structure, ready for action again in the knowledge that threats of this nature are always subject to revival. In 1994 another essentially defensive effort was also launched in the pages of *Home Power Magazine* to respond to growing threats that major investor-owned electric utilities might effectively take control of the home power industry.[2] Yet there appears to be little or no appetite for capitalizing on real successes, as traditional models might lead us to expect, to form permanent and growing membership organizations, launch agendas, or advance particular candidates in a political context reaching beyond the local community. Little or no effort appears to be made to tap the vigorously active and effective "power base" revealed in these defensive actions for other ends in the fashion of traditional party or electoral politics.

This is not to say that home power participants are not interested in the larger world. In my experience, they are uniformly and almost without exception eager to share their home power and other experience and to go well out of their way to assist others in this connection. The energy fairs, publishing, and other educational outreach efforts (including vigorous interest and participation in international development efforts) that characterize the movement indicate a far greater dedication to affecting the broad body politic than we would expect from the average citizen. Membership in environmental organizations such as the Sierra Club or Greenpeace is surely not uncommon either, although action may again remain locally centered. There appears, simply, to be little appetite for exercising the traditional levers of power in state or national politics as a way of pressing local perspectives and agendas on the nation at large. Fully, even fanatically, committed in their own efforts to live differently,[3] and eager to explain what they are about or aid others who may be interested, home power participants vigorously defend their efforts against subversion from the outside. But they appear to have little interest in making their practices and perspectives mandatory through the traditional channels of politics and public policy making. Such a course literally does not seem to come to mind, absorbed as they are in alternative efforts of their own and in assisting others who are interested.

This failure to pursue what might be called a traditional "political growth" model may be explained partly by a sense that there is not much prospect for success in any effort to change the nation as a whole. The uniformity of the prevailing pattern of absorption in local action, however, suggests that a better explanation may lie in the lessons of non-violence (again probably traceable to the civil rights, antiwar, and other movements of the 1960s), and in a simple, though by traditional norms surprising, lack of interest in the centralized exercise of power. However it is explained, the result is a genuine alternative politics of participation.

PARTICIPATION IN THE ECONOMY

The unique selectivity characterizing political participation extends also to home power's embrace of the traditional patterns of economic activity. Carl's small construction company, for example, remains distinct from the customary pattern in at least two ways, in spite of its recent incorporation and the inclusion of insurance and other benefits for its employees. For one thing, Gimme Shelter's customers continue to be local residents with strong extraeconomic commitments to energy conservation and environmental sustainability. They become closely involved in the design and construction process and generally participate directly in construction to some degree, coming to look more like cooperative participants in one of the company's projects from earlier years than like "customers" purchasing the "products" of corporate enterprise. As a second major departure from normal expectations, employees are not required to keep regular schedules but work instead only on those days they choose. As good friends, they do not abandon each other or fail to show up as promised, of course, but work cooperatively to match their collective efforts to the tasks at hand.

Traditional business models do not serve well in general, in fact. Small-scale installers, for example, report that they would

rather not have any employees, preferring instead to work cooperatively with those who are in market terms their competitors
but who they know share their own home power commitments.[4]
A number of dealers have quite deliberately chosen to remain
small, cutting their mailing lists or restricting their product line
to prevent growth beyond themselves[5] or themselves and one or
two neighbors.[6] Lacking clearly defined alternative models,
"business growth" has occasionally taken hold only to fail or be
rejected. In one case, a conglomerate (Photocomm) that bought
out a number of small dealers and, for a time, clearly set out on a
business course, found that ordinary employees could not sustain
the enterprise. After a severe downturn, this company in one instance relinquished one of the businesses it had purchased, simply abandoning it back to its original owner, who then rapidly
built it back to success. Learning from the experience, Photocomm's more recent acquisitions have involved little or no change
in personnel or practices in the businesses they have acquired.[7]
Those responsible for the Amherst energy fair have so far also
shied away from proposals for the purchase of a permanent site,
for shifting the fair closer to a major population center with
greater commercial potential, and for long-term funding links to
environmental education and other established interest groups,
in spite of the tremendous burdens of the present heavy reliance
on volunteer efforts centered in their own small community. In
these and other cases, "business" models have been embraced
only selectively, where they could contribute without significantly undermining primary goals.*

At the level of individual home power adopters there is further
evidence of departures from normal economic practice. This is
most obvious, of course, in the home power decision itself, given
the fact that photovoltaic power systems, for example, are widely
agreed to produce electricity at least twice as expensive as utility
power. Bank financing for home power homes is also quite rare.

---

*Having said all of this, one of the more interesting questions for the future
of home power efforts is the degree to which classical economic structures may
eventually capture the movement in the absence of well-developed organizational and practical alternatives. There is a firm if often inchoate resistance to
simple acquiescence to classical market formulations but it is not yet clear to
what degree this resistance will simply be worn away rather than contribute to
the emergence of well-defined and durable alternatives.

This may be due in part to the fact that traditional mortgages would in many cases not be obtainable even today. But in the vast majority of cases the people themselves have explicitly preferred to remain independent in their efforts and unencumbered by substantial debts.

Surprisingly, in a purely material sense many home power people appear to be as well off as or even better off than their more traditional middle-class counterparts. They typically own their homes outright and, as their home power systems have matured, have begun to share in many traditional middle-class comforts. A degree of purely material success may even have been assured, in some instances, by the specific determination to separate more fully from the mainstream economy and commit full-time to their own pursuits. Devoting full time to the cultivation of often markedly underutilized rural land and other resources, the traditional shift toward two full-time wage earners seems to have been reversed, with an increasing amount of productive time spent in and around the home. "Success" seems, for example, to have followed the termination of one couple's seasonal employment driving a truck—a job that removed both entirely from the work of developing their remote home site near Timbo, Arkansas, for a major portion of the year. A full-time commitment to a large garden, a milk cow, and the construction of a log home from materials on their land were combined in this instance with the restoration of a local building (now rented by a small bakery) and a part-time summer job performing music in a local restaurant. The nontraditional mix involving a wide range of productive activities has resulted in a very attractive home, which is entirely paid for, and a purely material standard of living for this couple in their forties and their two children that is essentially indistinguishable from the mainstream of the middle class.

The Amish in various parts of this country have, of course, long demonstrated that very different mixes of productive activity and consumption patterns are sustainable under a given set of material conditions. Individuals involved with home power appear in many instances to have found niches that lie, one could say, somewhere between Amish and mainstream patterns. Similar creative revisionings of what are ordinarily taken to be the "natural selection" criteria of economic life[8] may prove to be as essential to a democratic shaping of technology as the emergence of a new politics of participation.

KNOWLEDGE AND EXPERTISE

The politics of knowledge and expertise in the home power movement also depart from the patterns of mainstream culture. Knowledge tends to be relatively widely distributed, with the owner-installer, for example, generally able to talk directly with those formally involved in the design, manufacture, or sale of home power systems. "Experts," for the most part, are direct participants themselves, closely involved in developing their own home power homes. And design improvements continue to involve direct participation by users. This pattern has prevailed from the development of the first really reliable inverters (actually developed by principals in Trace Engineering who were fed up with their own experience with unreliable inverters on a sailboat) to the ongoing design of small electric vehicles through interactions at the MREA Energy Fair. It is also evident in the origins of what was in 1996 probably the fastest-selling wind machine on the market (Southwest Windpower's Air 303), which was specifically designed in response to home power experience and people's expressed desire for a lightweight machine that would cost less than five hundred dollars and not require a sophisticated tower. The Air 303 also responds to the popularity of the modular nature of PV panels and can itself be installed in multiple numbers like a "wind module."

The inclusiveness of the design process to date is undoubtedly due in part to the nature of the technology itself and to its stage of development. In the early days, much of the work amounted to a kind of backyard tinkering with already existing machinery (e.g., old wind machines) or components (e.g., automotive alternators adapted for use in micro-hydro units). Diverse electrical, mechanical, business, and other skills were required, without a high degree of specialization in any area, and dedicated individuals were able to assemble what knowledge was needed from books and through interactions among themselves. In the apparent absence of the kinds of market potential that might have brought the specialized design teams of major corporations into play, home power systems remained relatively simple from a technical standpoint.

Even as more sophisticated technical knowledge is brought increasingly to bear today, the business remains small enough, and the need for close connections to the end user strong enough, that the traditional gulf between users and technical experts still has not emerged. There also remains what appears to be an irreducible need for a closer awareness on the part of the home owner of his/her own energy use; at least until home power systems become so inexpensive as to be routinely oversized, it will not be possible for home owners simply to pick up additional appliances (especially things such as resistance space heaters!) at the store, come home, and plug them in. This need for awareness of energy use helps to ensure a level of knowledge well beyond that of the typical consumer of utility power and helps to maintain what remains a narrowed distinction between user and expert in home power. Grid-tied systems such as the roof-mounted PV arrays being installed by the Sacramento Municipal Utility District as part of their Solar Pioneers program may, on the other hand, remain essentially transparent to the home owner. As such, they could provide an avenue for a return to more traditional lay/expert relationships.

The currently narrowed distinction between user and expert also appears to be a product of intentional efforts that are probably linked to some of the motivating drives behind home power itself. A desire not for self-sufficiency but for greater "independence," for example, is widely reflected in efforts to learn and know more about home power systems than would be essential to their use under traditional models. One person in the Amherst community with little familiarity with electricity, for example, expressed strong regrets that he had not dropped his own construction efforts building an addition on his own home in order to follow more closely the installation of his home power system by someone else in the community. Not having done so, he remained more dependent on his installer (who spends winters in Florida) and on others in the community than he really wishes to be. There are signs, also, of an occasionally explicit effort simply to remain in closer touch with the world and with the fundamental relationships that are expressed through home power systems. Karen's log of sun and wind conditions is but one illustration of this enhanced appreciation of connections with the natural world. Motivating interests of this sort again tend to imply a more informed and involved user and, by extension, a narrowed gap between that user and others with greater expertise.

PUBLIC EDUCATION AND THE DISSEMINATION OF KNOWLEDGE

The lack of an appreciable gap between experts and users in home power is clearly related also to a fourth distinctive feature of home power that takes the form of a unique dedication to educational efforts and the dissemination of knowledge. This dedication is especially evident in the efforts of Richard and Karen Perez and their various coworkers at *Home Power Magazine*. Entirely devoted to the sharing of knowledge and experience in the home power movement, *Home Power Magazine* regularly features the stories of individual home owners, often in articles written by the home owners themselves. These stories range from that of a British law graduate who moved to a remote part of Scotland and completely fabricated his own collection of wind machines,[9] to the headmaster of a private school in Arizona whose daughter's experience at Arco Santi led him and his wife to build their own domed home, equipping it with PV power and building in their own top-loading freezer and refrigerator.[10] These stories along with other features ranging from "Basics of Alternating Current Electricity" to "Regenerative Braking a DC Series Motor" (articles in the same two issues of the magazine), are designed to be accessible to any interested layman. And they clearly reflect the Perezes's passionate belief that such information simply "should" be available and accessible to the general public.

Unique commitments to the dissemination of knowledge are also evident in energy fairs such as the one in Amherst, in the ordinary course of business among home power dealers who must often spend an extended period of time educating potential home power adopters, and in interactions with people who have adopted home power systems all across the country. This dedication to sharing knowledge is, again, necessitated in part by the nature of the technology and by the stage of its development as a business. Home power dealers who fail to educate their prospective customers regarding the importance of attending to efficient electricity use, for example, are not likely to end up with satisfied home power customers. At the same time, those interested in home power often arrive with a desire for the independence afforded by their own understanding of their systems and therefore tend to be eager students. In none of these areas, however, could the effort devoted to knowledge dissemination be considered even remotely cost-effective by traditional standards. A major burden added to energy systems that are not, themselves, economically competitive, it does not begin to pay for itself but rests, instead, on the personal commitments of participants in the movement.

The knowledge dissemination efforts of the home power movement stand in sharp contrast to traditional cultural patterns under which knowledge is often carefully if unconsciously husbanded as a source of income for the knowledge holder. This contrast is perhaps most evident in comparisons with recently emerging utility home power programs in which systems are designed and delivered with virtually no home-owner participation beyond an audit of electricity use. Typically placed in a locked container to which the customer has neither physical nor legal access, these systems come as close as possible to the monopolization of knowledge characteristic of traditional utility power systems in which the consumer's role is limited to the use of energy and regular payment of the utility's bill. The "locked box" approach may very well be justified by liability concerns on the part of utilities. Such a justification does nothing, however, to mitigate the disparity between customers and experts that it perpetuates.

8

A Theoretical Perspective

THERE IS, IN HOME POWER EFFORTS, A REMARKABLE IF LARGELY UN-
examined intelligence that has elicited from deep commitments
to the underlying values of democracy a fascinating model for
more democratic technology design and development. This model
resonates well with several bodies of theory that help define it
and offer added insight into its practical expression.

There is, for example, an immediate connection with a set of
conceptual foundations Lewis Mumford has described for what
he has called "democratic technics."

> [T]he spinal principle of democracy is to place what is common to all
> men above that which any organization, institution, or group may
> claim for itself. . . .
> Around this central principle clusters a group of related ideas and
> practices with a long foreground in history. . . . Among these items are
> communal self-government, free communication as between equals,
> unimpeded access to the common store of knowledge, protection
> against arbitrary external controls, and a sense of individual moral
> responsibility for behavior that affects the whole community.[1]

In the home power movement, we see a vigorous collective effort
for people to take charge of their own lives. We see a virtual elimi-
nation of "lay"/"expert" distinctions and the barriers they present to communication in the shaping of technology. We see an
almost fanatical dedication to educational efforts, creating a
uniquely accessible and widely shared common store of knowledge. We see a pattern of political involvement aimed specifically
at protection from arbitrary external interventions ranging from
the siting of a nuclear waste repository to the exploitation of local
environments. And, in the outspoken statements and vigorous action of home power people, we see a vigorous "sense of individual

140

moral responsibility for behavior that affects the whole community" that is sometimes explicit (e.g., Roy's statements in the last section of this chapter), sometimes firmly present in essentially aesthetic commitments.

It is a second vigorous theoretical resonance, however, that will provide a focus for the remainder of this chapter: Benjamin Barber's conceptualization of a "strong democracy." The first order of business will be a brief summary of Barber's strong-democracy framework in which action is seen to flow from "political talk," "politics as epistemology," and direct forms of "public seeing and public doing." This will be followed by a section making the links between home power and Barber's theory explicit. The chapter will then conclude with a section drawing connections between a few other theoretical treatments of participation in technology decision making and the home power experience.

BARBER'S "STRONG DEMOCRACY"*

Barber begins with a particular notion of the shape and thrust of politics. The political realm, he suggests, is marked by three prominent conditions. There is a necessity for (public) action. There is conflict or potential conflict over what that action should be. And there is an absence of private or independent grounds for judgment. In this framework, "the ultimate political problem is one of action, not Truth or even Justice in the abstract." For Barber, then, "To be political is to *have* to choose—and, what is worse, to have to choose under the worst possible circumstances, when the grounds of choice are not given a priori or by fiat or by pure knowledge (*episteme*). To be political is thus to be free with a vengeance—to be free in the unwelcome sense of being without guiding standards or determining norms yet under an ineluctable pressure to act, and to act with deliberation and responsibility as well."[2]

Notably, contemporary notions of the social construction of technology would clearly place us in the realm of the political Bar-

---

*It will be impossible to do justice to Barber's conceptual framework in the limited space available here. Readers who find the summary offered here too much condensed to be intelligible may find that they can skip the section entirely and still grasp much of the argument that follows.

ber has described here. Whether actively or by default, we must and we do, for example, choose the energy systems by which our material existence is maintained. Yet the values inherent in any particular choice, fossil, nuclear, renewable or otherwise, in whatever particular configuration, cannot be weighed objectively. There are no independent, a priori grounds for any specific choice in the shaping of technology and we are left, quintessentially, in Barber's realm of the political.

Having outlined the defining difficulty of the political realm, Barber then takes the position that public action is best chosen through a kind of participatory "strong democracy," a combination of "public seeing" and "public doing" that pivots on the twin hinges of "political talk" and "politics as epistemology."

By political talk, Barber means to refer to a universally participatory continuing conversation intended to serve as a primary medium for social and individual growth and development and intended to be concerned less with argumentation than with genuine communication. In political talk, "no voice is [to be] privileged, no position advantaged. . . . Every expression is [to be taken as] both legitimate and provisional, a proximate and temporary position of a consciousness in evolution."[3] Barber emphasizes that strong democratic talk "entails listening[4] no less than speaking; [that] it is affective as well as cognitive; and [that] its intentionalism draws it out of the domain of pure reflection into the world of action."[5] Genuine political talk serves not only to articulate interests and to persuade but to explore mutuality, establish affiliation, maintain autonomy, and embrace witnessing and self-expression.[6] A conversation that "places its agenda at the center rather than at the beginning of its politics," political talk in a strong democracy must be inclusive, scrutinizing even "what remains unspoken, looking into the crevices of silence for signs of an unarticulated problem, a speechless victim, or a mute protester."[7]

In a sense, political talk "is not [simply] talk *about* the world [but] talk that makes and remakes the world"[8] as participants both enter the conversation themselves and reflect on and react to the contributions of others. In this sense, political talk is an essential element of Barber's second pivotal concept of "politics as epistemology." Under this concept, Barber argues that a strong democracy does not act on "knowledge" in any simple, scientific sense. Rather, knowledge must be taken to be those characterizations of reality democratically adopted to serve as the effective basis for action. From a classical scientific perspective, much of this "knowledge" might be regarded as mere belief or intuition or even dismissed as untrue. But in a strong democracy, knowledge in the sense of a basis for action must be determined by democratic, over and above scientific or other standards. Faced with concerns about global warming, for example, scientific standards of proof must to some degree be set aside. A political choice must be made between a continuation of customary energy practices with their associated carbon dioxide emissions and altering those practices and emissions. The knowledge base implicitly embraced in the selection of a course of action should, in Barber's framework, be arrived at democratically through the interactions of political talk. Science as epistemology may well be a significant element in the interactions of political talk, but it should be (democratic) politics as epistemology that ultimately governs action.

In admittedly dense summary form, Barber offers the following characterizations of politics as epistemology.

> When politics in the participatory mode becomes the source of political knowledge—when such knowledge is severed from formal philosophy and becomes its own epistemology—then knowledge itself is redefined in terms of the chief virtues of democratic politics. Where politics describes a sovereign realm, political knowledge is autonomous and independent of abstract grounds. Where politics describes a realm of action, political knowledge is applied or practical and can be portrayed as praxis. Where politics concerns itself with evolving consciousness and historically changing circumstances, political knowledge is provisional and flexible over time. Where politics is understood as the product of human artifice and contrivance, political knowledge is creative and willed—something made rather than something derived or represented. And, finally, where politics is the preeminent domain of things public (*res publica*), political knowledge is communal and consensual rather than either subjective (the product of private senses or of private reason) or objective (existing independently of individual wills).[9]

In Barber's framework, politics as epistemology situates us in the moment of what might be called the "social construction" of knowledge as a foundation for action. And it seeks to make that construction democratic. The imprimatur, "knowledge," is placed through a process that is not narrowly scientific but *democratic*, broadly participatory and concerned at least as much with action (i.e., "What shall we do?") as with "truth" in any more abstract sense.

Finally, in its reliance on political talk, strong democracy rests not only on "public seeing" in the sense of "politics as epistemology" but just as importantly on "public doing." "Democracy is neither government by the majority nor representative rule: it is citizen self-government."[10]

> In place of the search for a prepolitical independent ground or for an immutable rational plan, strong democracy relies on participation in an evolving problem-solving community that creates public ends where there were none before by means of its own activity and of its own existence as a focal point of the quest for mutual solutions. In such communities, public ends are neither extrapolated from absolutes nor "discovered" in a preexisting "hidden consensus." They are literally forged through the act of public participation, created through common deliberation and common action and the effect that deliberation and action have on interests, which change shape and direction when subjected to these participatory processes.[11]

"Public doing" is as essential in this process as is "public seeing" or public knowing.

For Barber, it is through participation in political talk and the ongoing accumulation of experience through action that knowledge as a basis for choosing among alternative actions is politically (democratically) constructed and, in resolving the political necessity of choice, enacted.

HOME POWER AS AN INSTANCE OF STRONG DEMOCRACY

In many ways, the home power movement is perhaps best seen as an expression of Barber's strong democracy. It traces its origins back to what might be described as the commitments to political talk of the countercultural movements of the 1960s. It steps outside authoritative images of proper energy and other practices and does not accept the traditional pronouncements of scientific or other expertise, shifting instead into a form of politics as epistemology. And it carries participation far beyond traditional expectations with respect to consumer behavior, complementing "public seeing" with vigorous forms of "public doing."

It can be argued, of course, that the extreme openness of the countercultural upheavals of the 1960s could not have been achieved without the influence of drugs and that normal inhibitions do not permit the range of thought and attitude expressed and accepted ("Do your own thing") in that time. It was not a period whose openness came without costs either, even if those costs have largely disappeared from view today, both in the home power movement and among those now successful in more traditional patterns of life. In its reaction to earlier times and to the experience of Vietnam, and in its occasional nihilism and self-indulgence, the period of the 1960s may well have been excessive and destructive, ultimately failing in fundamental ways to conform to Barber's ideal. Yet there was in that period an effective and remarkable commitment to something like Barber's political talk that created unique openings for consideration of alternative technologies, alternative values, and alternative patterns of life. In this sense, the home power movement can be said genuinely to have emerged from the positive and creative potential of political talk.

If it is viewed in isolation today, the home power movement may still appear to operate from an agenda that pulls it somewhat away from Barber's ideal of political talk. If one is attentive, however, there are many signs that commitments Barber would applaud remain quite real. These signs are sometimes subtle. One often hears, for example, expressions of regret that some of the interactions and free-ranging conversations that were once common are now more limited. A pioneering dealer and installer in Arkansas has expressed disappointment that the close interactions he and other young people moving to the hill country of the Ozarks once had with lifelong residents in the area no longer occur as they once did. There is regret expressed also in the Amherst home power community that the kinds of talk that were common among young people as they first came to that area have become more difficult as work and family responsibilities have intervened. Signs of this sort are sometimes more explicit. One woman, for example, specifically reports that one of the things she likes best about having left her job in New York City to build her own home power home in a small town in Idaho is that she now comes into direct contact with loggers and others whose environmental views differ sharply from her own. As a final illustration of subtle signs of an embrace of political talk, there is, in settings ranging from informal conversation to formal writing, a surprisingly widespread resistance to any tendency for one person to speak for another. Words are chosen carefully with an insistence, for example, on expressing a motivating interest in "independence" as opposed to "self-sufficiency" in the pursuit of home power systems, and the most innocently conceived effort to assist others in expressing themselves seems to be resisted in favor of the greater patience of a more careful listening attitude.

The movement's continuing commitment to political talk is, of course, more strikingly evident in its unique dedication to participatory research and education and in its selective approach to traditional political action: the dedication to offer perspectives and participate in the conversation is not accompanied by efforts to exclude or dominate others through the mechanisms of traditional electoral politics.

"Politics as epistemology," like political talk, has been vigorously embraced in the home power movement in the sense that the "knowledge" that serves as a foundation for action is politically defined rather than dependent upon "scientific" or other authoritative certification. Authoritative knowledge is, in fact, regularly controverted as, for example, PV systems are adopted in spite of their higher costs or wind systems are added to complement PV arrays in sites where neither produces a financial return because of relatively poor resource conditions for both sun and wind. Status in the knowledge community is assigned directly on the basis of one's ability to make systems work and essentially ignores the traditional signals of educational background or institutional affiliation. The mix that results among presenters, for example, at an energy fair, may make those with traditional technical backgrounds cringe at times. But action is not governed by such things as the notion that global warming, for example, has not yet been demonstrated beyond all scientific doubt. Relying on politics as epistemology, as those in the home power movement have, people act individually and collectively from their own best determinations. "Knowledge" is what they decide to "go with" as a basis for action, whether or not it has yet been validated by receiving the imprimatur of "scientific" or other authority.

This is not, of course, to say that home power people have rejected traditional science or engineering or that they are obstinately committed to perpetual motion or to the reinvention of the wheel. Rather, the function and direction of technical understandings have been altered. Where water pumping using less than a quarter as much energy may be "impossible" in the traditional framework, technical understandings can be redirected. Where electric "automobiles" prove infeasible with present battery technology, the use and performance characteristics of automobiles may be subject to renegotiations favorable to very differently configured electric "vehicles." Where photovoltaic power is found to be more expensive per unit of energy than utility power, "affordability" and "attractiveness" may prove to be more relevant standards. Narrowly scientific positions on environmental and other issues do not necessarily govern action. "Knowing" is not restricted to science, and the question is not simply how people can best fit in with scientific or technological "facts," but how science and technology can best serve actively and politically defined human aims. The human and political elements of knowledge, in other words, are recognized and embraced unflinchingly in practice.

In ways that resonate well with Barber's notions of public seeing and public doing, it is clear also that the home power movement has begun with participation and only arrived at the technology now going into place in the course of that participation. In ways that are consistent with the distinctive patterns of participation in traditional politics described earlier, this has not been a case of "participation in technology decision making" or an instance in which a particular group has decided to become involved, per se, in choosing their own energy technologies. A commitment to independent thought and a broad engagement in choosing carefully how to live have come first. The technologies of home power, along with other departures from tradition in work and community life, have followed.

Albert Wurth has been critical of traditional notions of public participation in technology decision making:

> The very framing of the question—public participation in technological decisions—reveals the conventional ordering or priority of the

components of the problem. The obvious temporal priority of the technological decision over the subsequent citizen participation accurately reflects the value priority of the parts of the decision process. The *decision* is technological; the *participation*, though perhaps more than an afterthought, is, at best, an agreeable supplement to the process. The decisions and the decision makers are technical, the participation and the role of the public are ancillary.[12]

Home power technologies, as movement participants often point out on their own initiative, are not the focus of participants' interest. There are, of course, those few who are indeed very fascinated with technical things. But on the whole, it is the technologies that are in this case "ancillary," mere means in the service of broader ends. In this sense, though surely not by carefully analyzed intent, participants in the home power movement have broken completely from traditional images of participation in technology decision making, placing full participation first and technology a distant second. We see evidence here not only of "political talk" and "politics as epistemology" but of a full-fledged instance of "public seeing" and "public doing."

PARTICIPATION AND MORAL AUTONOMY

Among the features of the home power experience that give it a uniquely democratic character both in the context of Barber's "strong democracy" and in certain other theoretical frameworks is the remarkable vitality of direct citizen participation. This final section in a chapter on theoretical perspectives will focus on distinctions between "pluralist" and "direct participationist" notions of participation and on notions of the emergence of "moral autonomy," with the object of further sharpening theoretical interpretations of participation in the home power movement and its link to a more democratic shaping of technology.

In his analysis of various forms of public participation in technology decision making, Frank Laird[13] has made an effort to derive evaluation criteria from two major theories of democracy, "pluralism" and "direct participation." He begins by outlining similarities and differences between these two theories. Pluralists, he notes are concerned with the participation of groups, while direct participationists are concerned with participation by individuals as individuals. Pluralists are concerned with outcomes in terms of the eventual distribution of burdens and benefits in a society, while direct participationists are as much concerned with the educational and psychological effects of participation on citizens as they are with outcomes. Pluralists also tend to assume that people's desires and interests arrive from outside the political process and remain fixed through that process, while direct participationists see democratic participation as an important factor in forming and changing the outlooks and attitudes of participants. For direct participationists, truly democratic participation "makes people more aware of the linkages between public and private interests, helps them develop a sense of justice, and is a critical part of the process of developing a sense of community." Effects of this sort are thought to be "more important than the actual political or policy outcome of any specific controversy."[14]

By the standards that Laird then derives for each theory, home power experience would be more likely to be applauded by direct participationists than by pluralists. Home power has done relatively little to bring new groups into the traditional (pluralist) policy process and has not focused on "group learning" through the attraction of expert personnel who might better articulate and press for group interests in Congress or other large forums in which the pluralist competition among interest groups takes place. Home power activities have done little to enhance access to, or extend influence over, traditional policy-making officials.

By the standards Laird offers under the theory of direct participation, however, home power has done relatively well. Home power continues to bring growing numbers of individuals into participation as individuals in ways that help both to educate them and to enhance their influence over the energy decisions that shape their lives. Home power experience also carries well beyond the dissemination of information to include the improvement of citizens' understanding. It helps to ease resource inequalities among those who might seek to shape energy decisions, principally through its dedication to free communication and a broad dissemination of knowledge regarding alternative energy systems. And it has not amounted to a process of delegation: citizens actually retain or gain substantial authority to make decisions and to act on them directly.

Noting that both pluralists and direct participationists see a crucial role for knowledge in technology decision making, Laird suggests that what is really needed in a democracy is what he calls "participatory analysis." Among the hallmarks of participatory analysis, he suggests, is participants' ability at some level to begin to analyze the problem at hand. It is essential that participants be "able to challenge the formulation of the problem itself" and "decide for themselves what the most important questions are." And it is essential that participants "structure their relationships to experts in such a way as to avoid losing their democratic prerogatives." To qualify as "participatory analysis," participants must somehow "avoid being taken in or co-opted by expert opinion."[15] Earlier discussion of the distinctive patterns of the home power movement suggests that it meets all of these standards for "participatory analysis" in exemplary fashion.

The nature of participation in the home power movement, however, appears to carry one step further in advancing the cause of democracy and a more democratic shaping of technology, reaching into the realm of what Zimmerman[16] and others before him have referred to as "moral autonomy." Zimmerman draws from the work of David Cooper in addressing the concept:

> Cooper sees morally autonomous persons as independent, not only "in the sense that the rules they follow are self-imposed," but also "in the sense that they are choosing with a full understanding of the public nature of moral rules" (1993, 119). Such individuals are deemed fully conscious of their connectedness to a public existence and their duty to follow moral principles with "a universal social purpose" (p. 120). This does not mean that such individuals all value the same things, but rather that they consider the broadest possible meaning and consequences of their choices.[17]

Participation in the home power movement has yet to reach a scale at which group participation might begin to supplant participation by individuals. Participation in the movement therefore necessarily implies the sort of exercise of independence entailed in home power's sharp departure from traditional societal norms. Individual histories and the comparative sacrifice of material self-interests that accompanies the home power choice appear also to distinguish the further sense of connectedness to universal social purpose that is indicative of moral autonomy.

Among occasionally outspoken participants there are clear declarations of independence from mainstream thought marked by a profound sense of moral responsibility that has been carried beyond words into independent action. After many years of vigorous activity as a home power pioneer, Roy has put the matter in direct and simple terms:

> What we have in the United States for sure, and to a certain extent the whole world, is not a technology problem. We've got all kinds of technology. But our problem is attitude. [W]e refuse to acknowledge what's really going on. . . . We can't do what we're doing. It's not . . . you know, socially responsible is one way of putting it. But plain old, simple "consciencable" [sic]—or you know the word I'm looking for. [Interviewer: "Unconscionable?"] OK, there you are. It is [unconscionable]. If our ancestors would have done to the earth what we're doing, first of all, we might not be here. But secondly we would curse them. And we would say that this was the worst generation to ever live on the planet. (Or several generations.) Because we ain't payin' the bill. We're trying to *cheat*. And it's not fair. So I'm a big thing on attitude. You know the reality is, too, we can't wait for the government to do this, to force us to do it. . . . We can't wait for it to happen from the top down. It's got to be from the bottom up. And it's not a revolution that's violent. It's a revolution of attitude. But we got to change it.[18]

It is impossible to say, of course, to what degree participation in technology decision making might on the one hand have contributed to the development of moral autonomy and, on the other hand, to what degree the latter might have led to the former. If we adopt the view of direct participationists, however, the benefits of moral autonomy for democracy would, in either case, extend well beyond any practical benefit that might accrue from the development of alternative sources of energy. Ultimately, the moral autonomy that Zimmerman describes, and many home power people have shown in practice, may be indispensable to a more democratic shaping of technology.

9

The Challenge of Home Power: Toward a More Democratic Shaping of Technology

In a formal sense, we commonly acknowledge the fact that democracy is not a process that can be sustained on autopilot. We often refer, for example, to Benjamin Franklin's classic response during the Constitutional Convention to questions about what kind of government the founders were devising for the new nation: "A republic; if you can keep it," he is reported to have said. Yet as we make the practical technological choices that so significantly shape our lives, we often act as if democracy were an automatic process.

We must come to recognize in our conduct that political talk, politics as epistemology, public seeing and public doing are all as essential in the design and development of technology as they are in any matter of public consequence.

HOME POWER AS A MODEL

As a model for more democratic technology design and development, home power offers us at least two demonstrations that may be of great value. The first of these is not developed in detail here but may be worth mentioning: home power experience provides a rare and concrete illustration of apparently viable technological alternatives that are expressive of values distinct from those incorporated in more conventional patterns of technology.[1] While home power may not be seen as a complete solution to all of our problems in urban, international, and other contexts, it does provide a far more complete image of the potential for departures from traditional patterns than might easily be imagined in any simple substitution of one energy form for another. There are, in other words, real choices to be made. Technical and economic considerations do not bring us apolitically to single-valued technological imperatives.[2]

As a second, equally valuable contribution, home power has firmly demonstrated the possibility of a design and development process built from its foundations on direct public participation. This demonstration is especially robust and its significance particularly far-reaching in that it has occurred in an area (energy production and use) in which basic comprehension is ordinarily taken to require a level of technical sophistication well beyond that of the ordinary citizen. We know from the home power movement, in other words, that it is possible to develop a process that shifts the normal strictures of economic rationality and of political feasibility to a secondary role, embraces a free exchange of information, and breaks down lay/expert distinctions in such a way that ordinary people effectively define as well as participate directly in the resolution of their own problems in areas bearing on technology.

An activated population would appear to be the central defining feature of the home power model. In this case, activation, a sense of problem ownership, and a kind of recovery of agency, have not been restricted to one narrow area of concern but seem to have been extended to a continuing engagement with every significant aspect of life. Although some degree of specialization has survived and people are not simply going it alone, one gets the feeling that nothing has been entirely delegated. In the final analysis, people are in no instance fully abandoning their independent capacities.

Along with this activation has come a degree of implicit if not explicit wisdom—an independent compass or even a degree of moral autonomy, if you will—that has guided participants in the home power movement not only to a range of new technological developments but to an altered participation in politics and the economy and to a unique commitment to "unimpeded access to the common store of knowledge."

Technology has arisen within this context not as an issue in itself or in isolation from a more general engagement. And expertise has remained relatively narrowly circumscribed, functioning within the community and in service of that community. Experts, to the degree that they can be distinguished from the lay population, have not been set apart or positioned to dominate or to set the terms of discussion. They have, in other words, remained "on tap," not "on top."[3] Continued dedication to this pattern both on the part of adopters and of many home power businesses suggests, at some level, a conscious or conscientious dedication to "independence" as an alternative to our more customary tendency toward dependence on experts of professional (e.g., legal and medical), corporate, government, educational, and every other stripe.

As a model opening the horizon to new practices in the shaping of technology, home power offers a challenge to the patterns of the past. Its challenge to the governance of science and technology by relatively closed circles of expert peers in the tradition of Vannevar Bush's *Science: The Endless Frontier*[4] is direct and perhaps self-evident. But it raises questions also about classical policy-making frameworks and even about the kinds of developments in participatory technology assessment that have been seen recently in Europe.

In very practical terms, this challenge might lead us to ask what would happen if the widespread NIMBYism* of recent years were to mature along lines suggested by the home power movement; supposing participation in the form of simple opposition to particular technologies were to shift toward comparably vigorous public participation in the design and development of new initiatives. Prospects of this nature could turn out to challenge our very definition of democracy. We could find ourselves asking not how to encourage greater participation but whether we should be concerned about too much participation.

With a summary image of the home power model in mind, this and other prominent challenges the model may pose for traditional thought and practice will now be explored in greater detail. The next section will place the model in the context of recent developments in citizen-oriented technology assessment efforts, suggesting that these are not yet fully satisfactory. We will then return to a further consideration of NIMBYism and use this as a foundation for confronting the issue of the desirability of participation directly. The chapter closes with a section outlining specific measures that might be taken to encourage shifts in the

---

*"Not In My Backyard" opposition to technology.

direction of the home power model, should this be regarded as a desirable direction for change.

A CONTRAST WITH CLASSICAL AND NEWLY EMERGING POLICY PERSPECTIVES

To the degree that we traditionally assume that science and technology are best left to scientists and engineers, the home power model offers a challenge to custom. Where the locus of problem definition and solution might ordinarily be restricted to a relatively closed circle of experts (corporate, government, and university) and the population at large might be expected to be brought into conformity with authoritative findings through public education and appropriately designed legal and market incentives, home power seems to have violated all of the usual expectations. In doing so, it may prove to deal more successfully with energy, environmental, and many other less recognized difficulties than the best of what the expert community has had to offer to date.

Without doubt, the participatory processes of home power developments have at least helped to overcome the widespread paralysis in which relatively passive "consumers" of science and technology remain disconnected from the realities that may urgently call for changes in their patterns of behavior, while "experts" seemingly charged only with providing the means to continue observed (hence, deductively, preferred) behavior see themselves as avoiding any meddling in people's "choice of lifestyle." Judged by instrumental standards alone, lodging problem definition and solution with ordinary people cannot easily be dismissed as an alternative to common practice, even in relatively technical spheres such as energy production and use. Adding the other citizenship and democratic "process" benefits of participation, a fundamental reconsideration of conventional approaches to the governance of science and technology may be in order.

If we were to be convinced that a more democratic design and development process actually would be broadly desirable, our thinking might still be dominated by classical policy perspectives in such a way that we would remain tempted to ask, "How do we get people to behave more like this?" Looking for ways to "make it happen," we might consider legislative action, incentive systems, the formation of new organizations or institutional structures, or increased support for a wide range of educational and public information programs—all to "get" people to participate.

In some measure, these are the kinds of responses we already see in the "consensus conference" and other "participatory technology assessment" efforts currently flourishing in a number of European countries[5] though, notably, not to the same degree in the U.S.[6] In some of these cases, ordinary citizens are formally impaneled in place of the usual technical experts or representatives of major stakeholder groups and asked for their own lay assessments of particular technological prospects. They may be paid in the way that citizen jurors are paid in court proceedings. And they may be assisted in their deliberations by explanatory presentations from a variety of technical experts.

Efforts of this sort are assuredly improvements both on technology assessment procedures that lack provisions for direct citizen involvement and, more obviously, on older forms of technology decision making in which the perspectives and interests of scientists and engineers may have prevailed without serious examination or opportunity for public challenge.

The home power model, however, challenges us to go well beyond any recognized form of participatory technology assessment. Most significantly, perhaps, it does not begin from a set of relatively specific technological proposals. It doesn't even begin from a technically defined field of concern such as "energy production" or "environmental pollution." It begins, instead, from the lives of ordinary citizens. Full participation comes before technology. Technology arises as an issue for attention only in the ordinary course of a full engagement with life in all its aspects. What technology is taken up for examination—indeed, whether technology becomes an issue to address at all—is a question that remains very much on the agenda. Under this alternative model, new technologies are less prone to "come at" the population from such sources as advances in genetic engineering or a particular corporation's concerns with global competitiveness, for example. Instead they can be expected to emerge more directly from popular desires such as an interest in a sustaining sense of community or a more meaningful work life—or, of course, from the more familiar desire to improve material standards of living.

Even "science shops" as they now offer research services to the public in Denmark and as they are being developed in this country[7] do not quite meet the standard of the home power model. Again, an admirable step in the right direction, even science shops retain aspects of the tradition of citizen or layman as "supplicant," applying to a center of knowledge or "higher learning" for assistance, which may or may not be forthcoming depending upon the appropriateness of the request in the context of standards of inquiry set perhaps less by citizens than by the center itself, in combination with other centers like it. It is one thing for universities and other major institutions to throw open their doors, allowing entry to those often excluded in the past. Under the home power model, however, those open doors might more significantly permit those already inside to make an exit and become, themselves, "supplicants" seeking to be of more direct service again to the society that ultimately supports them. Under the home power model (and in Barber's strong democracy), knowledge is defined by society through the political interactions of democracy and it is up to every citizen, technically trained or otherwise, to bring his/her wares into the conversation for general consideration; it is not the citizen's burden to apply for assistance at centers of knowledge sanctioned by some supposedly higher authority.

This is not, of course, to deny the significance of our best understandings of material possibility at any given time. It is only practically and formally to recover the sense that these understandings are to be used in the service of society, not as a foundation for privileged positions of power from which society may be shaped through technology in ways that may ultimately run contrary to the defining principles of a democracy.

A FUTURE FOR NIMBYISM?

Within a traditional policy context, it is still conceivable that the question of how we might get the shaping of technology to occur through a more democratic design and development process could prove to miss the mark entirely. Efforts to encourage greater participation in technology decision making through "technology literacy" programs, and even through such innovations as science shops and participatory technology assessment, have admittedly met with only very limited success to date. Yet claims to the prerogative of deciding matters of technology as they arise in ordinary life have reached what some would regard as epidemic proportions[8] in the form of "NIMBY" opposition to everything from waste incinerators and nuclear waste repositories to high-voltage power lines and facilities for basic research in biochemistry. The willingness on the part of citizens simply to say, "No," to authoritatively promoted technologies, even when those very technologies may quite obviously be materially essential in their own day-to-day lives (e.g., power lines), continues in striking contrast to apparent nonparticipation in other settings.

In some respects, this seeming heavy-handedness on the part of ordinary citizens in one setting and failure to make a showing in another may directly reflect precisely the kinds of distinctions that have been made above between "participation in technology decision making" and technology decision making that arises in the ordinary course of a more broadly defined "participation." If, in any case, the home power model should prove a good outline of patterns toward which NIMBYism as a whole may now be evolving, it could turn out that the proper policy question will be more one of how to handle a sea change in modern practice than one of how to nurture and encourage greater participation.

Not surprisingly, NIMBYism tends to provoke consternation on the part of traditionally recognized expert communities and a response whose sardonic questions are not far below the surface: "Well, if you don't want this, if you reject the best advice of those most knowledgeable in the relevant fields, where is your better idea? (If you're so smart,) what do you suggest?" To some degree, this response may put NIMBYs back on their heels for a time. Often entirely dependent, in the short run, on the very technologies they oppose, their position may indeed seem rationally indefensible. And they are not generally equipped or prepared to propose alternatives, often only having grasped rather recently the unacceptability of a particular action involving technology and having as yet little comprehended the way in which that technology is already interwoven in an elaborate fabric upon which they are themselves so dependent. If, however, they should begin turning the experts' response itself around, their complaints might increasingly take the character of a challenge: *"You* claim to be the expert, it's not *my* job to come up with alternative means. That's at least partly *your* job. What you have proposed so far is unacceptable and it's up to you to work *with* me to see how else we might proceed." Moving from simple opposition into the kinds of direct participation in a fundamental reformulation of problems and exploration of solutions that has characterized the home power movement could sharply alter the nature of NIMBY activities. Rather than working to get more participation along the lines of the home power model, our task might turn out to be simply coping with the currents of change implied by whole-sale shifts in this direction.[9]

Is Real Participation Such a Good Idea?

Contemplating this prospect, there may yet be reason to pause and ask whether more democratic design and development of technology actually is desirable. Even if we are inclined to ap-plaud certain achievements of the home power movement itself, we are clearly dealing in this case with a very select group of citi-zens. These are survivors of a grueling series of trials that have assuredly denied most people the opportunity of participation in the careful shaping of technology. The vast majority even of those who moved to the countryside back in the 1960s assuredly are not among those now putting the finishing touches on their own home power homes, having fallen by the wayside long ago or been forced by financial and other exigencies back into the fold of mainstream technological practice. The vast majority of us do not normally participate as full partners (citizens) in the design and development of technology but choose instead only as "consum-ers" of technology from a set of market options profoundly nar-rowed in comparison with the far broader realm of material and sociocultural possibility.

Before we jump into more democratic technology design and development with both feet, we may wish to ask exactly how many people we can expect to be both able and willing to partici-pate in the way that home power people have. If provisions for more open participation somehow could be made, exactly who and how many might participate? And what kind of technology— what sort of a reshaping of our lives—might we see as a result?

Images of a world not anchored as ours is by the knowledge structures of scientific expertise can indeed be rather disconcerting. Do we wish to risk the balkanization not only of technology but of society and culture in ways that could contribute to serious divisions and even conflict? Do we wish to validate the expression through technology of everything from Scientology to the militia movement?

In the case of home power, there is already clear evidence of a failure on the part of the popular media to grasp the sense of home power activity. Where it is noted at all, it is depicted only as evidence of falling prices and an approach to passing the classical (but in fact largely irrelevant) tests of cost competitiveness.[10] Or it is painted as a subset of the survivalist wing of the militia movement[11] when the overlap between the two is demonstrably vanishingly small.[12] Home power itself seems to be accompanied by an increase in civic spiritedness and, if anything, a broadly sharing rather than a narrowly aggressive program of proselytization.[13] As the faulty media images indicate, however, the prospect for balkanization and a breakdown of communications is real—and need not remain in all cases as benign as it has with home power.

What, moreover, about the simple matter of material error? As we elevate lay knowledge and "politics as epistemology," don't we increase the risk that global warming or some other scientifically discernible catastrophe might overwhelm an increasingly disrespectful and disbelieving public? Should we not, in fact, be vigorously defending and upholding the authority of science out of a basic concern for our own collective survival?

In the end what we come down to here is the simple question of our commitment to, and definition of, "democracy." This is not a question the founders themselves dealt with unequivocally. Still, it might be argued that democracy's inspiration lies in the idea that if we ask and expect much of the ordinary citizen, we will on average both elevate the individual and improve upon the collective prospect. By contrast, operating from the assumption that individuals will act only in their own short-term material self-interest (and that they cannot in general be expected even to identify where that interest lies without expert assistance) might be taken as a recipe for decay in the individual and for despotism in the body politic.

Historical and other arguments can obviously be made that democratic systems are more successful—or less successful—by material or by other standards than other systems.[14] These arguments can, in turn, be elaborated to support or undermine any particular tradition or institutionalization of democracy. Similar arguments can be made with specific reference to technology and technology decision making.

At some point, commitments to a more democratic process for shaping technology probably rest on commitments to democracy as an article of faith and on a willingness at least to risk material penalities in the name of the values of democracy.

Still, a recognition of some of the hazards just outlined may be helpful in minimizing the downside and maximizing the benefits of a more democratic handling of technology decision making. Such a recognition may also enhance appreciation for the kinds of moral autonomy as a benefit of participation that were described at the end of the previous chapter.

The Challenge—Moving Toward a More Democratic Shaping of Technology

The principal task in moving toward more democratic technology design and development along the lines of the home power model will be the challenge of rejoining a sundered conversation. This may, in turn, require much more in the way of removing obstacles to improved practices than of formal measures to get people to behave differently.

If we were to take as the ideal of democratically responsive design the situation in which all factors relevant to design were universally shared—from the most sophisticated understandings of material possibility to a full appreciation of value and other human commitments—then the closest approach to this ideal would rest on the best achievable communication among those who inevitably hold some but not all of these factors. The conversation (or to use Barber's term, the "political talk") underlying such communication is one of the most critical things that has been restored in the case of the home power movement. And we can identify from home power experience at least three objectives likely to contribute to a generally improved conversation.

- We need to bring scientific and technical understandings back into the popular exchange of ordinary life.

- We need to work toward a new technology and technological agenda in which we not only attend to, but act from, an awareness of the human and socio-political, as well as material and instrumental, effects of technological choices.

- We need to cultivate modes of exchange in which different values and objectives can surface and interact with technical understandings in ways that will contribute to the reconstitution of all three.

The first of these objectives will require some adjustment in traditional priorities. Technology, we must recall, does or does not happen *within* democracy, not the other way around. And it is citizens, collectively, who ultimately rule this land: we must remind ourselves that our Constitution does not establish or empower experts as final arbiters charged with the reduction of politics to a series of narrowly technical comparisons to be drawn and acted upon by those few individuals best credentialed in what they themselves suppose to be the relevant fields. The ultimate task of the technically trained is service not to corporate or otherwise limited interests but to society and to the nation and world as a whole.[15] This task necessarily implies contributions to the development of a range of technological alternatives, not single technologies already interwoven in a dependency on many others with one unelaborated choice, "take it or leave it," for the whole.

There will be a need here to bring technical people back into the lay community in ways that will loosen the shared images of the world and definitions of problems that characterize their traditionally insular disciplinary homes. Engineering students at our elite universities, for example, need to be taken out of the labs in which they are often essentially free to "play with their toys," to come into direct confrontation with joblessness and with settings such as inner-city minority communities. They need to be able to place their tinkerings with things such as advanced computer networks or electric automobiles in context and at least reflect with others in society on the degree to which their efforts do or do not respond to pressing human needs. Scientists and engineers, generally, need to be removed at times from their corporate, university, and government settings so that they can become sensitive to the possibility of goal displacement—i.e., instances in which particular technical goals become institutionalized and continue to be pursued even when they may fail to serve (or begin actively to undermine) the public interest.

Aggregate economic interests are demonstrably not identical to the public interest; unserved elements of the latter that are not reflected in the former have a tendency to fester until they emerge as major problems in society, with environmental concerns being only one well-recognized case in point. Much might be done to avoid such problems if the actual design practices of members of the scientific and engineering community could be brought into lay settings. Rather than having "public hearings" collecting lay reactions to well-formed technical proposals—i.e., public participation in technically framed proceedings—the design process itself would be moved into a popular setting governed at least in part by its lay participation.[16]

A more democratic design and development process is likely to lead to different technology. Smaller-scale, less technically sophisticated, more easily understood technology could well emerge if designs were to flow increasingly from the general population rather than the more limited participation of those with formal technical training.

Deliberate efforts could also be made by the government and others, however, to encourage greater accessibility in technological devices and a reversal of our common tendency to move increasingly in the direction of devices that conceal the mechanisms by which they do their work.[17] Efforts of this sort might encourage the recovery of basic competencies that tend to be lost under our system of specialized interdependency. Without necessarily requiring greater attention from users, technologies that invite entry or at least make understanding possible, rather than encouraging a "switch flipping," "know-nothing" attitude, could build rather than undermine the independent capacities of citizens with respect to technology. This need not mean further formal education in all cases even along the lines of present "technology literacy" programs. In many instances essential technical understandings are not so esoteric. Technical people genuinely concerned with serving the public and with an unbiased clarification of what is and is not materially possible could contribute greatly to the transparency of the operating principles incorporated in ordinary technology. And we might eventually expect much more from lay participation, given the different experience of exposure to such alternative technology.

Certain negative effects of large-scale technology may be worthy of special attention as targets in themselves. The bureaucratic subdivision of agency (i.e., of the capacity to act independently) typically associated with the large-scale, for example, may undermine individual responsibility and effectively incapacitate both those who rely on, and those who construct and operate, technologies in this category. With individual actions constrained to a pattern that is designed to ensure efficient operation of the whole, "agency" is in large measure delegated to that pattern, and any that may remain unclaimed has been subdivided to the point of atomization. Abstract and authoritative knowledge structures become almost necessary for the whole to function in a coordinated manner. Specialization also tends to undermine conversation, as the narrow focus and specialized language of separate disciplines compartmentalizes interests and excludes not only laymen but other specialists. Political talk and politics as epistemology have been virtually extinguished in this situation, and "public doing" from these sources is similarly difficult to imagine. The "political" has essentially been replaced (for as long as the enterprise can be sustained) by the "planned."[18]

The history and demise of nuclear power development in this country could be interpreted as a popular rejection of technology that was simply too large in scale and too inaccessible to citizens—too impregnable to citizen entry, influence, or participation—to be acceptable. More democratic design and development efforts might well embrace and act from this message. This is not to say that there will never be instances in which material efficiencies or other considerations might outweigh other possible effects on users or on society. But these effects, and perhaps especially the effects of the large-scale, need to be given real weight as they are balanced among traditional concerns in the process of shaping technology.

Ultimately, we will need to cultivate modes of exchange in which different values and objectives can surface and interact with technical understandings in ways that will contribute to the reconstitution of all three. Efforts to bring technical understandings back into ordinary exchange will contribute to this third objective. So will moves toward a new technology more attentive to human and social effects. Science shops and the various modes of participatory technology assessment currently under development should also make a contribution. Beyond these, however, there may be a need to work explicitly to counterbalance some of the biases that are inherent in traditional corporate, government, and university-based research and development programs. New organizational forms or a significant realignment of incentives may prove to be appropriate as a way of tempering economic and other forces that have perhaps become too pervasive and all-encompassing in our society. Direct efforts may be called for to shift research and development resources back toward popular roots.

As has been indicated, home power has emerged for the most part not with the assistance or support of traditional institutional forms but in a clearing relatively free from their direct influence. If we wish to nurture a more democratic process for the design and development of technology, it may be appropriate to actively establish and protect "clearings" of this sort rather than simply hope for the kind of fortuitous opening that has been associated with the home power movement.

A mismatch of institutional forms is, in fact, widely observable at many points in the home power movement today. Traditional business growth models clearly are not appropriate in the minds of many home power dealers or for many of those involved in the Midwest Renewable Energy Association. Yet dealers who trim their mailing lists and work deliberately to remain small may eventually be eclipsed by big business competitors. If the MREA does not acquire a permanent site, routinize its funding by becoming involved in regular environmental education or other programs, or at least move its annual fair to a more populous and commercially attractive site, its long-term survival also may be at risk.

Active experimentation with intentional community arrangements (often involving land trusts) and with softened corporate forms (e.g., Carl's construction company's loosened work hours and strengthened involvement with its home-building "customers") is widespread. Few well-developed alternative models are available, however, and it may well be appropriate to consider assistance in this area. In fact, depending upon the value one places on a more democratic shaping of technology, and on the kind of close interactions essential to such a shaping, outright protections may be in order to preserve, for example, smaller-scale home power businesses against utility-scale reorganization.

The values and commitments of the home power movement go well beyond the boundaries of conventional political, economic, and technical discourse. In the absence of the unique circumstances of their emergence, it is difficult to imagine either the reconceptualization of material possibility or the concrete development of technological alternatives that now gives them expression. A revival of voluntary associations, an expansion of the nonprofit sector, or other organizational and institutional changes may naturally enhance opportunities for the kinds of exchange that make a more democratic design and development of technology possible. But there may well be room for the active cultivation of such alternatives, as well, in an effort to incorporate the democratic shaping of technology as a matter of commonly accepted social practice. At some point, a conscious dismantling and a conscientious reconstruction of certain meta-technological systems—i.e., infrastructural and other technological systems that fundamentally shape the functioning of a technological society—could prove to be a useful approach.[19]

Closing

As a model for a more democratic shaping of technology, the experience of the home power movement has the potential for augmenting traditional practice in important ways. The degree to which it might be appropriate for this model to supplant other approaches to technology decision making remains a question open to discussion and debate (i.e., a suitable topic for "political talk"). Among those who find a more direct democratic practice appealing, however, the movement offers many possible lessons and implications.

The first and most important lesson, perhaps, is that it is possible to shape (design and develop) technology through a directly participatory process governed by ordinary people. The technological outcomes, while they may be vastly different from traditional outcomes, may be promising not only materially but in terms of their innovative recombinations of material possibility with human values and commitments. Beyond this, it appears that an activated citizenry and a close interaction of traditionally separated lay and expert knowledge may be essential to a more democratic process. And it may be necessary to create or protect the kind of organizationally and institutionally open terrain within which the home power movement has flourished in order to allow room for practical and theoretical innovation and experimentation in the development process.

A shift in the center of gravity in the practice of technology decision making toward the home power model would very likely result in a different mix of technologies in society. Simpler, more accessible, smaller in scale, this technology might well produce somewhat less in a material sense or operate somewhat less efficiently by traditional measures than the systems that would otherwise prevail. This sort of trade-off, itself, would likely be a subject for much political talk. Such technological alternatives, however, might also be more accessible to tinkering and might do more to invite and encourage technical competence on the part of citizens, tending in turn to cultivate the kinds of democratic participation in the shaping of technology that brought them into being in the first place.

If we wish to encourage developments along the lines of the home power model, we must think in terms of participation first and technology second. It may also be useful to work, through public information or other programs, to counter classical images of science and engineering as rarefied pursuits properly isolated from ordinary human affairs. Efforts to bring traditional experts into the field and to cultivate previously sundered conversations wherever possible may be the most urgent need. But it is also possible that NIMBY and other developments will naturally evolve in the direction of the home power model and that our chief task could prove more to be keeping the way clear and the evolution orderly than one of getting people to enter more vigorously into the fray.

Under the banner of "progress," new technology has long been either eagerly embraced or resignedly accepted in the Western world. Whether by choice or by default, that technology has come in large measure to define us as human beings.

In an important sense we [have] become the beings who work on assembly lines, who talk on telephones, who do our figuring on pocket calculators, who eat processed foods, who clean our homes with powerful chemicals.[20]

In historical terms, it is interesting to ponder exactly how the technology we depend upon—the technology that in large measure defines us—has come into being. How was it selected? Who selected it? And what alternatives got rejected along the way? How, that is, have we come to be who we are?

There is yet a more important question that arises with respect to the future, however: "If this is who we are, who else could we be?" In a democracy, this is a question we should all have every opportunity to be involved with directly.

10

A Brief Epilogue

THE TWO STUDIES PRESENTED IN THIS VOLUME ARE IN MANY WAYS very different, and yet there are strong similarities. In many respects, in fact, the home power movement may seem to flow in an entirely logical evolutionary line from the values and commitments evident among members of the Alliance. Left at the end of Part I with questions about what the world Alliance members would like to live in might look like, the home power movement appears to carry us a long way toward a clearer image of precisely such a world. While members of the Alliance seemed to have very little prospect for putting their values into practice, the continuing developments of the home power movement have vastly improved those prospects.

Comparing the earlier experience of the Alliance with the more recent history of the home power movement, one may also be struck by the differences in context. Members of the Alliance were virtually forced into an almost entirely defensive stance, confronted as they were by the saber rattling of the Reagan years, the continuing development and start-up of new nuclear power plants, the environmental assaults of James Watt, and the other vigorous threats they saw to a desirable pattern of life. Whether the tides have turned only temporarily with changes in national and international politics or whether shifts such as the fall of the Berlin Wall and the (temporary?) demise of nuclear power will prove more lasting remains to be determined.

The alternative views represented both in the Alliance and in the home power movement do, in any case, appear to be enjoying further development in the interim. Together, these studies do not begin to provide us with a complete historical picture, of course. Yet they may justify some hope that continuing developments could begin to provide a foundation for multiple images of our technological future—images rich, realistic, and attractive enough that they could, themselves, stimulate discussion and more direct participation in defining and choosing that future.

If the promise of choice, which is the most important promise of the home power movement, is to be realized—indeed, if the promise of democracy and of technology itself is to be realized—we will have to learn to listen more carefully and more successfully, especially to the more muted voices among us. Whether they are muted by disposition, because they are not yet fully developed (still struggling with the as yet inchoate), or by virtue of restrictions from the outside, they must be heard. For the promise of the future to be realized, especially with respect to the shaping of technology, we must learn to "listen" well. Again,

> "I will listen" means to the strong democrat not that I will scan my adversary's position for weaknesses and potential trade-offs, nor even (as a minimalist might think) that I will tolerantly permit him to say whatever he chooses. It means, rather, "I will put myself in his place, I will try to understand."[1]

This, after all, is the spirit of fair play, of a respect for human dignity—of democracy at its best.

Notes

INTRODUCTION

1. Bruno Latour, *Science in Action* (Cambridge: Harvard University Press, 1987).

2. Wiebe E. Bijker, Thomas P. Hughes, and Trevor Pinch, *The Social Construction of Technological Systems* (Cambridge: MIT Press, 1987).

3. The term "political construction" was first suggested to me by Norman Vig, professor of political science at Carleton College, in a personal communication in 1992.

4. For a theoretical development of this issue from a human rights perspective, see Jesse S. Tatum, "Technology and Liberty: Enriching the Conversation," *Technology In Society* 18, no. 1 (1996): 41–59.

5. Benjamin Barber, *Strong Democracy: Participatory Politics for a New Age* (Berkeley, Calif.: University of California Press, 1984).

6. Ibid., 182.

7. For a profoundly useful discussion of the "promise of technology" see Albert Borgmann, *Technology and the Character of Contemporary Life* (Chicago: University of Chicago Press, 1984). (Quoted phrases taken from page 36.)

8. Langdon Winner, *The Whale and the Reactor* (Chicago: University of Chicago Press, 1986), 12.

9. Barber, *Strong Democracy*, 175.

10. Berkeley anthropology professor Laura Nader gives eloquent expression to her struggles with this difficulty during her work with the National Academy of Sciences' Committee on Nuclear and Alternative Energy Systems in "Barriers to Thinking New About Energy," *Physics Today* (February 1981), 98–104.

11. David J. Rose, "Continuity and Change: Thinking in New Ways About Large and Persistent Problems," *Technology Review* (February/March 1981), 53–67.

12. I have elsewhere explored several arguments from basic human rights to the effect that individuals should in some instances enjoy protection from certain technological developments. See Tatum, "Technology and Liberty: Enriching the Conversation."

13. Barber, *Strong Democracy*.

CHAPTER 1: THE ALLIANCE AS A GROUP

1. Initial contacts with the particular group that is the subject of this work were established in the spring of 1981. Sporadic contact was maintained between the spring of 1981 and the spring of 1983 based solely on the author's personal curiosity regarding an unfamiliar pattern of life. The formal study re-

ported here was initiated early in the spring of 1983 and involved more regular contact with the group over a period of a year, including several months of actual residence in the group's community. A draft of this study was first completed in October 1985. Limited contacts in the fall of 1987 indicated that the group was continuing to function essentially as it had earlier, still maintaining its rented office and with six of the seven members described in chapter 2 continuing to participate in the group essentially as they had in 1984. As presented here, however, this study describes the state of affairs before 1985.

2. Peter Schwartz (manager, Future Studies, SRI International), "Luncheon Address," in *Proceedings of the Third Stationary Source Combustion Symposium*, vol. 5, addendum (February 1979), EPA–600/7–79–050a.

3. Ibid.

4. Amory Lovins, *Soft Energy Paths: Toward a Durable Peace* (Cambridge, Mass.: Ballinger, 1977). Herman E. Daly, ed., *Economics, Ecology, Ethics: Essays Toward a Steady-State Economy* (San Francisco: W.H. Freeman, 1980).

5. Schwartz, "Luncheon Address;" Lovins, *Soft Energy Paths*; Daly, *Economics, Ecology, Ethics*.

6. Duane Elgin and Arnold Mitchell, "Voluntary Simplicity (3)," *Coevolutionary Quarterly* (summer 1977), 4–19. Duane Elgin, *Voluntary Simplicity: Toward a Way of Life That is Outwardly Simple, Inwardly Rich* (New York: William Morrow, 1981).

7. Willis Harman et al., "Broader Issues," chap. 5 in *Solar Energy in America's Future: A Preliminary Assessment* (prepared by Stanford Research Institute International for the Energy Research and Development Administration, March 1977, DSE–115/1), 81.

8. Ibid., 90.

9. Ibid.

10. Ibid., 97.

11. Ibid.

12. Robert Coles, *Children of Crisis* (Boston: Little, Brown, 1967).

13. Michael Agar, *The Professional Stranger: An Informal Introduction to Ethnography* (New York: Academic Press, 1980).

14. The term "community" is used to refer to the three towns from whose populations virtually all participants in the Alliance were drawn.

15. Despite the very limited income and assets of its active members, one member of the group (referred to in later sections by the name "Peter") was able to obtain unsecured loans from individual members amounting to several thousand dollars. Inability to repay those loans as promised greatly strained his relationship with the group; the trust implied in making the loans is, however, exemplary of the kind of relationship that existed among members of the Alliance.

16. The title refers to the blue line actually painted on the pavement outside the Diablo Canyon main entry gate. Crossing this line constitutes trespassing and makes protesters subject to arrest. More will be said about the play in a later section on its author, "Bill."

17. This activity was actually initiated by the local Women's Party for Survival and cosponsored by the Alliance. See later discussion of "activities."

18. A former Alliance member living in another state worked for this networking organization.

19. See, for example, Steven E. Barkan, "Strategic, Tactical and Organizational Dilemmas of the Protest Movement Against Nuclear Power," *Social Problems* 27, no. 1 (October 1979).

20. Meetings began at 7:30 p.m. and this rescheduling of topics often occurred during the course of the meeting.

21. While Bill's personal support during the eight or ten months of the play effort was provided in substantial part by family and friends outside the geographic community, this support remained a small fraction of the total he and others collected for the play from within the community.

CHAPTER 2: ALLIANCE MEMBERS AS INDIVIDUALS

1. It may be important to some readers to note that Peter was the only member of the Alliance who was, to the author's knowledge, a regular drug user. In Peter's case, marijuana use has been observed as frequently as three times (at intervals of several hours) in the course of one day. Drugs, including alcohol, beer, and wine—and even, in conformance with some member's definitions, sugar and caffeine—were almost entirely absent in the author's observations of individuals and of group meetings and events, with only two other exceptions: (1) at a birthday party held by the group for one of its active members, a small glass or two of wine, each, was consumed by a few participants and (2) at an outdoor gathering hosted by the Alliance for a group of about thirty people passing through on a transcontinental "walk for the earth," a few members of the Alliance joined in some marijuana offered around the circled gathering by the walkers.

CHAPTER 3: INTERPRETATION AND SIGNIFICANCE

1. The term "internalized" is employed here as it is defined and used by Elliot Aronson: "The internalization of a value or belief is the most permanent, most deeply rooted response to social influence. The motivation to internalize a particular belief is the desire to be right. Thus the reward for the belief is intrinsic. . . . Once it is part of our own system, it . . . will become extremely resistant to change." Aronson contrasts conformity to group norms based on "compliance" (accompanied by external rewards and/or punishments) and on "identification" (the desire to "be like" another individual) with that based on "internalized" values or beliefs. See Aronson, *The Social Animal*, 3d ed. (San Francisco: W.H. Freeman, 1980), 30.

2. Note, for example, Bill's comments regarding his experience with law school or Peter's remark regarding the adult "models" in his early life: "They kept telling me I was sick because I was feeling something different."

3. As Peter put it: "The speed is too fast. . . . It's like . . . 'How you doing?' and they walk on by without waiting for an answer, you know? That's always thrown me for a loop."

4. This distinction between useful/satisfying work and work that pays well is apparently drawn from more than the direct personal experiences that have already been described for Brad, Bill, and others. Murphy, for example, observes that his father has always been a general practitioner rather than a specialist partly because "he likes to just help people." (At age 71, Murphy's father had begun consulting at a weight loss clinic but continued to concentrate his efforts on his general practice because "that's what he likes" despite much higher eco-

nomic incentives at the weight loss clinic.) As another example, Peter observed that making money is not generally "just a twist of fate. It's all planned, methodical. And most of the people who've made their money have not made it from righteous community-serving ideas and endeavors; they've made it off of somebody else—stepping on somebody else." Expressing a contrasting ideal, Peter continued: "I mean, to operate . . . from the point of view that everybody has to win in every encounter or relationship is something that's . . . not a commonly shared perspective on business."

5. For a cogent outline of the dominant "commodity" approach to the analysis of energy policy, along with three potentially competing alternative perspectives and a discussion of some of the limitations of conventional views, see Paul C. Stern and Elliot Aronson, eds., *Energy Use: The Human Dimension* (New York: W.H. Freeman, 1984). To quote a brief sample here, the dominant "commodity view" takes energy to be simply

> a commodity or, more accurately, a collection of commodities. Energy means electricity, coal, oil, and natural gas. . . . When people talk about "U.S. energy supplies" or "projected energy demand," they are usually talking about this list of tradeable goods. Commodity energy consists of energy forms or energy sources that can be developed and sold to consumers. . . .
>
> The view of energy as a commodity reflects a certain set of values and beliefs; acting on this view tends to move particular interests to the center of attention. The commodity view emphasizes the value of choice for present-day consumers and producers. It assumes that such choice will allocate energy (and other commodities) effectively and efficiently. It also assumes that when prices rise, fuel substitutes will be found and that inequities that arise can be handled by ad hoc modifications to the system. It focuses analysis on the transaction between buyer and seller and away from other aspects of energy use that are external to the transaction. The interest of energy producers, along with those of consumers who have sufficient resources to participate in energy markets, take center stage. The effects of energy use on environmental values, social equity, occupational and public health, the international balance of payments, and the like are considered secondary, and people who are concerned with such effects must petition the political system for attention to those issues. (15–17)

Stern and Aronson's discussion of three other widely held, though politically secondary, views moves a little toward the kind of departure we see among members of the Alliance. They discuss views of energy as an "ecological resource" (renewable or non-renewable, polluting or nonpolluting, etc.), as a "social necessity" (e.g., for cooking and heating), and as a "strategic material" (with unique national security and geopolitical significance). Again, however, "In most aspects of the national policy process, the commodity view is dominant. Dominance of a particular view of energy does not mean that it is the only view given consideration, but that other views must make special claims before being taken seriously. And in most U.S. energy policy debates, the burden of proof still remains on those who assert that energy should be treated as something other than an ordinary commodity" (23). See also, Jesse S. Tatum, *Energy Possibilities: Rethinking Alternatives and the Choice Making Process* (Albany, N.Y.: State University of New York Press, 1995).

6. Steven Lukes, *Power: A Radical View* (London: Macmillan, 1974).

7. Parallel arguments have been made in the context of Appalachian poverty in John Gaventa, *Power and Powerlessness: Quiescence and Rebellion in an Appalachian Valley* (Urbana, Ill.: University of Illinois Press, 1980).

8. John D. McCarthy and Mayer N. Zald, "Resource Mobilization and Social

Movements: A Partial Theory," *American Journal of Sociology* 82, no. 6 (1977): 1214.

9. Neil Smelser, *Theory of Collective Behavior* (New York: The Free Press, Macmillan, 1962).

10. T. R. Gurr, *Why Men Rebel* (Princeton: Princeton University Press, 1970).

11. R. N. Turner and L. Killian, *Collective Behavior*, 2d ed. (Englewood Cliffs, N.J.: Prentice-Hall, 1972).

12. McCarthy and Zald, "Resource Mobilization."

13. Barkan, "Strategic, Tactical, and Organizational Dilemmas."

14. Smelser, *Theory of Collective Behavior*. Although some would see the references to the theory of collective behavior made here as somewhat out of date (the study is presented essentially as it was written in the mid 1980s), I believe the older theory in this field continues to offer greater insight into the observed behavior of the Alliance and its members than more recent theoretical developments.

15. Ibid., 8.

16. Smelser refers to this, in an analogy to economic production processes, as a "value added" process—each determinant contributing to the definition of a particular collective behavior outcome.

17. Ibid., 313.

18. Ibid., 120. Internal quotation is from C. Kluckhohn, "Values and Value-Orientations," in *Toward a General Theory of Action*, p. 411. See Smelser, p. 25.

19. Dorothy Leonard-Barton and Everett M. Rogers, "Voluntary Simplicity in California: Precursor or Fad?" (paper presented at the annual meeting of the American Association for the Advancement of Science, San Francisco, 7 January 1980) [Palo Alto, Calif.: Institute for Communication Research, Stanford University]; Elgin and Mitchell, "Voluntary Simplicity (3)" Elgin, *Voluntary Simplicity: Toward a Way of Life That is Outwardly Simple, Inwardly Rich*.

20. Richard W. Scott, *Organizations: Rational, Natural, and Open Systems* (Englewood Cliffs, N.J.: Prentice-Hall, 1981).

21. A. F. C. Wallace, "Revitalization Movements," *American Anthropologist* 58 (1956): 264–81.

22. Ibid., 268.

23. Smelser, *Theory of Collective Behavior*, 364–65.

24. David J. Rose, "Continuity and Change: Thinking in New Ways about Large and Persistent Problems," *Technology Review* (February/March 1981), 54.

25. Bruno Bettelheim, *The Informed Heart: Autonomy in a Mass Age* (New York: Avon Books, 1960).

CHAPTER 4: THE HOME POWER MOVEMENT

1. Jesse S. Tatum, "Political Construction of Technology: A Call for Constructive Technology Assessment," *Research in Philosophy and Technology* 15 (1995), 103–15.

2. Ronald N. Giere, "Science and Technology Studies: Prospects for an Enlightened Postmodern Synthesis," *Science, Technology, and Human Values* 18, no. 1 (winter 1993): 102–12.

3. This apprehension is best conveyed by the burgeoning literature in the

field of science, technology, and society. Two excellent points of entry to this literature with this message clearly presented would be Winner's book, *The Whale and the Reactor*, and Randall Stross's anthology, *Technology and Society in Twentieth Century America* (Belmont, Calif.: Wadsworth, 1989). With specific reference to how the shaping of our lives through technology might legitimately need to be more carefully restricted than it traditionally has been, see Tatum, "Technology and Liberty: Enriching the Conversation," 41–59.

4. Vannevar Bush, *Science: The Endless Frontier* (Washington D.C.: National Science Foundation, 1945; reprint, 1960).

5. See, for example, the remarks of Congressman George Brown in C. Cordes, "As Chairman of Key House Committee Restates His Vision, Scientists Worry," *Chronicle of Higher Education*, 8 September 1993, sec. A, p. 26.

6. Significant recent work would include: Daniel J. Fiorino, "Citizen Participation and Environmental Risk: A Survey of Institutional Mechanisms," *Science, Technology, and Human Values* 15, no. 2 (spring 1990): 226–43; Philip J. Frankenfeld, "Technological Citizenship: A Normative Framework for Risk Studies," *Science, Technology, and Human Values* 17, no. 4 (autumn 1992): 459–84; Frank N. Laird, "Participatory Analysis, Democracy, and Technological Decision Making," *Science, Technology, and Human Values* 18, no. 3 (summer 1993): 341–61; Andrew D. Zimmerman, "Toward a More Democratic Ethic of Technological Governance," *Science, Technology, and Human Values* 20, no. 1 (winter 1995): 86–107; and Richard Sclove, *Democracy and Technology* (New York: Guilford Press, 1995).

7. At this writing, little has been published in English on these developments. Richard Sclove's book (ibid.) contains some discussion, and Norman Vig offers a summary of developments in "Parliamentary Technology Assessment in Europe: Comparative Evolution" (paper presented at the annual meeting of the American Political Science Association, Washington, D.C., 3–6 September 1992). A good collection of essays describing experience in several countries appears to be forthcoming from Vig's 1996 NSF study "Multivisioning the Future: Parliamentary Technology Assessment in Europe" (National Science Foundation grant SBR–9421908).

8. Phil Brown, "When the Public Knows Better: Popular Epidemiology Challenges the System," *Environment* 35, no. 8 (October 1993), 16–20, 32–41.

9. Langdon Winner, "Citizen Virtues in a Technological Order," *Inquiry* 35 (1992), 341–61.

10. I have made more extensive efforts in this direction elsewhere. See Tatum, *Energy Possibilities* or Tatum, "Technology and Values: Getting Beyond the 'Device Paradigm' Impasse," *Science, Technology, and Human Values* 19, no. 1 (winter 1994): 70–87.

11. Barber, *Strong Democracy*.

12. NSF-sponsored work has included extended residence and interviewing in the area of Amherst, Wisconsin, and approximately eight weeks of travel all over the United States interviewing home power businesses and home owners, as well as representatives of *Home Power Magazine*, municipal and investor-owned utilities, organizations such as Independent Power Producers, and others involved in the home power movement.

13. Earlier publications from this research include: Jesse S. Tatum, "The Home Power Movement: Technology, Behavior, and the Environment," in *Proceedings of the 1990 Summer Study of the American Council for an Energy Efficient Economy*, vol. 2, *Human Dimensions* (Washington, D.C.: American Council for an Energy-Efficient Economy, 1990), 141–49; Tatum, "The Home Power

Movement and the Assumptions of Energy Policy Analysis," *Energy—The International Journal*, 17, no. 2 (February 1992): 99–108; Tatum, "Technology and Values"; and Tatum, *Energy Possibilities*.

14. The estimate of one hundred thousand home power homes in the United States appears in an article in *Time* magazine ("Here Comes the Sun," 18 October 1993, p. 84) and is also the estimate offered by John Schaeffer, president of Real Goods Trading Corp., one of the largest retailers in the business. Paul Maycock, editor of *Photovoltaic News* (PV Energy Systems, 8536 Greenwich Road, Catlett, Virginia 22019) offers a contemporaneous estimate of sixty thousand homes, arguing that the higher estimate probably includes a number of summer cabins and other less than complete "homes," but he agrees that one hundred thousand is also a believable count (personal communication with author, 9 May 1996). Maycock is a former director of photovoltaic programs at the Department of Energy and continues to be one of the closest observers of photovoltaic developments.

15. This is Maycock's number (ibid.), but it is also consistent with my own observations and best estimates.

16. Sun Selector now offers inverter, charge controller, circuit breakers, and metering in a single box that needs only connections to batteries and PV panels. See *Home Power Magazine* 52 (1996), 23. British Petroleum Solar offers a packaged wall-mounted system that adds a microprocessor to inverter, breakers, metering, and other components, allowing centralized metering and control of many grid-connected home power systems. See *Home Power Magazine* 52 (1996), 33.

17. From this group, interviews have been conducted in 1996 at Jade Mountain (Colorado), Sunelco (Montana), Alternative Energy Engineering (California), Photocomm (multiple locations), Solar Energy Specialties (California), Hitney Solar Products (Arizona), and Real Goods Trading Corp. (California, Wisconsin, and Oregon).

18. Two of the oldest and most successful home power businesses (Real Goods and Jade Mountain) can, in fact, be linked back to a small rural food cooperative in California. Although precise statistics are obviously not available, it has been suggested that the bare-footed marijuana growers who came to town with their jeans pockets full of cash probably played an important role in underwriting the early development of home power as a business.

19. This story also appears in written form, for example, in the Jade Mountain catalog, fall/winter 1995/96, p. 111. One major figure half-jokingly reports that the whole home power movement is the fault of the Grateful Dead; if he and others had not been determined to play their Grateful Dead recordings, it might never have happened.

20. Photovoltaic cells themselves, of course, were originally developed as a part of the space program.

21. Tatum, "The Home Power Movement," and Tatum, "Technology and Values."

22. One indicator of developing interests in this direction is the publication of Gene Logsdon's book *The Contrary Farmer* by Real Goods Trading Corp. (1995) as one of its ten or so books for the home power market.

23. Examples of such innovations can be found at "Narrow Ridge," north of Knoxville, Tennessee, at the Yoga-based intentional community where the very popular Ananda Power Center (a prepackaged set of home power components) is manufactured and in the present form of "The Farm," a former commune in Tennessee. For formal discussion of the last of these, see Albert Bates, "Techno-

logical Innovation in a Rural Intentional Community, 1971–1987," *Bulletin of Science, Technology, and Society* 8 (1988): 183–99.

24. Among the most significant of these nonprofits are Solar Energy International of Carbondale, Colorado, and the Solar Electric Light Fund of Washington, D.C.

25. One of the more interesting approaches has been the bolt-on electric drive for a standard bicycle offered for approximately five hundred dollars by ZAP Power Systems of Sebastopol, California, both directly and through Real Goods.

26. Tatum, "The Home Power Movement."

27. Jade Mountain catalog, fall/winter 1995/96, p. 111. This catalog statement continues, "The outlook at Jade Mountain is to support and assist those who are striving to live appropriately on the planet by choosing tools accurate [sic] to a situation, tools that do the job and no more. We want to foster a feeling of life in harmony, a view of right livelihood which is environmentally friendly. . . . The vision of Jade Mountain is the fulfillment of appropriate technology— less is more."

28. Interview by author, Santa Fe, New Mexico, 21 March 1996.

29. Interview by author, Dreyfus, Kentucky, 8 March 1996.

CHAPTER 5: HOME POWER AS PARTICIPATORY RESEARCH

1. I take this term from work such as Peter Park, M. Brydon-Miller, B. Hall, and T. Jackson, eds., *Voices of Change: Participatory Research in the United States and Canada* (Westport, Conn.: Bergin & Garvey, 1993).

2. Mark Klein, James McKnight, Ray Reser, and Dave Shatz, "Masonry Stoves," *Home Power Magazine* 51 (February/March 1996), 42–46.

CHAPTER 6: ORIGINS AND EXPLANATIONS

1. This practice of proceeding without plans has been explicitly and repeatedly recognized in interviews in the Amherst, Wisconsin, area, in North Fork, California, near Ashland, Oregon, near Fox, Arkansas, and in many other home power situations across the country.

2. Interview by author, near Redstar, Arkansas, March 1996.

3. One often hears from people involved in home power that they don't "deal well with the negativity of it" (referring to opposition to classical energy policies) but prefer to do the "positive stuff" (interview by author near Dreyfus, Kentucky, March 1996) or that they prefer to work on the positive side, developing a positive energy agenda while others (e.g., one Midwest Renewable Energy Fair informant's brother, June 1995) carry on the effort to defeat what they regard as undesirable agendas. In general, the focus on protest and opposition of the 1960s has been left far behind as the "activation" of the present is fully and powerfully engaged in its own positive agenda. It is rare even to hear comments such as that of another Amherst informant, reporting that the first Amherst energy fair stemmed in part from the observation that "the fed" was unlikely ever to do anything about the environment, no matter what party was in power, and that this was part of the stimulus for members of the Amherst

community to begin thinking of doing their own educational fair. Negative comments about federal or other levels of action are rare and these traditional foci are essentially of little or no consequence as the positive agendas of experiment and development have become a consuming avocation.

4. The more complete statement from which this sentence is taken nicely reflects both the disjunctures of the past and the positive attitudes toward the future that have come to displace them:

> It's my opinion, prejudice maybe, that we [baby boomers] particularly have this attitude that, you know, it can be solved, whatever it is. It's not too scary to confront. And I don't know if that comes from, you know, my experience, for example, back in the late '60s of traveling in caravans to Washington, D.C. to march in the capital and being confronted by helmeted police and . . . realizing that I didn't have to be afraid of that. You know, if you can face something like that. Or my friends who went to Vietnam, or whatever. If you can face those kinds of situations where the fear, you would think, would overcome you. And then . . . maybe see [these] as difficulties and whatnot but find your way through them and still be able to stand and say, "This is what I believe." I think that gives you a certain edge, you know, in dealing with problems like this. . . . At least it does for me, I think, in that I kind of revel in the idea of alternative solutions, you know. Even if there isn't one readily available, I have a general belief that either there is one somewhere or we can make one. And I think that's an aspect of American character that I really love about us as a people. Lots of things we—I mean we've missed the boat on. But that, I think, is something that is very valuable and can carry us through—and *maybe* it's not just American, I don't know. Maybe it's worldwide, you know. . . . So it's not like . . . we're the only ones in the world doing this stuff. But it tends to be, I think, folks doing this who have an attitude that, whatever the problem is, it can be handled. It can be looked at, appreciated, and then solved. . . . I think that basically we have begun to believe that we really can design our lives, you know? (Interview by author, near Flagstaff, Arizona, March 1996)

CHAPTER 7: DISTINCTIVE PATTERNS

1. Myles Horton of the Highlander Research and Education Center (New Market, Tennessee) has suggested that this may be a pattern endemic even to the major educational institutions of mainstream society (personal communication with author, Berkeley, Calif., 1983.)

2. An organization now known as Independent Power Producers was launched in part for this purpose in late 1993. See Don Loweburg and Bob-O Schultze, "The Utilities, the Offgrid Market and IPP," *Home Power Magazine* 38 (December 1993/January 1994). Since this initial article, *Home Power Magazine* has featured a regular column from IPP.

3. In an apt turn of phrase, Roy has taken note of the large number of people in the Amherst, Wisconsin, area who seek, as he puts it, to "live carefully."

4. Sentiments of this sort have been made explicit in conversations with small-scale home power designers and installers in Wisconsin, New Mexico, Idaho, Vermont, and elsewhere across the country.

5. Windy Dankoff has deliberately restricted his core business to water-pumping systems, partly in order to remain small (Interview by author with Windy Dankoff, Santa Fe, New Mexico, 21 March 1996.)

6. Backwoods Solar reports that it has deliberately trimmed its mailing list for this purpose (Interview by author with Steve and Elizabeth Willey, Sand Point, Idaho, 15 April 1996.)

7. This description is based on conversations with several current and former employees of Photocomm.

8. See David F. Noble, *Forces of Production* (New York: Alfred A. Knopf, 1984), 144–45.

9. *Home Power Magazine* 52 (1996), 20–22.

10. *Home Power Magazine* 38 (1994), 6–10.

CHAPTER 8: A THEORETICAL PERSPECTIVE

1. Lewis Mumford, "Authoritarian and Democratic Technics," *Technology and Culture: An Anthology*, ed. Melvin Kranzberg and William H. Davenport (New York: Schocken Books, 1972), 50–59.

2. Benjamin Barber, *Strong Democracy,* 120–21. (Barber's emphasis.)

3. Ibid., 183.

4. Ibid., 175.

5. Ibid., 174.

6. Each of these is, of course, elaborated by Barber in his book. With respect to "maintaining autonomy," for example, he suggests that "[t]alk is the principal mechanism by which we can retest and thus repossess our convictions, which means that a democracy that does not institutionalize talk will soon be without autonomous citizens, though men and women who call themselves citizens may from time to time deliberate, choose, and vote." (Ibid., 190.)

7. Ibid., 182. Barber's discussion of political talk closely approaches Paul Feyerabend's call in *Science in a Free Society* (London: Verso Editions/NLB, 1978, p. 85) for an "open exchange" as opposed to a restrictedly "rational exchange." Barber writes: "Democratic politics cannot assume a paradigmatic language that is rooted in prepolitical syntax because it is itself *about* paradigmatic language." (Barber's italics, Barber, *Strong Democracy*, 157.) Albert Borgmann's notion of "deictic discourse" also comes close to Barber's political talk. See Borgmann, *Technology and the Character of Contemporary Life.*

8. Barber, *Strong Democracy*, 177. (Barber's italics.)

9. Ibid., 167.

10. Ibid., 211.

11. Ibid., 151–52.

12. Albert H. Wurth, "Public Participation in Technological Decisions: A New Model," *Bulletin of Science, Technology, and Society* 12 (1992): 289–93. (Wurth's emphasis.)

13. Laird, "Participatory Analysis, Democracy, and Technological Decision Making," 341–61.

14. Ibid., 345.

15. Ibid., 354.

16. Zimmerman, "Toward a More Democratic Ethic of Technological Governance," 86–107.

17. Ibid., 95. (Zimmerman's parenthetical references are to David Cooper, *Value Pluralism and Ethical Choice*, New York: St. Martin's Press, 1993.)

18. Interview by author, Amherst, Wisconsin, 28 June 1995.

CHAPTER 9: THE CHALLENGE OF HOME POWER: TOWARD A
MORE DEMOCRATIC SHAPING OF TECHNOLOGY

1. The value of home power as an illustration of another way to live in the world is developed at greater length in Tatum, *Energy Possibilities*.

2. This point has also been made in many places in the literature on science, technology, and society. With particular reference to supposed economic determination, see the work of David Noble, including *Forces of Production* (New York: Alfred A. Knopf, 1984).

3. I do not know who originated the phrase "experts on tap, not on top." I first heard it used by David Mathews, president of the Kettering Foundation (Dayton, Ohio) at the American Civic Forum Conference at the Hall of States, Washington, D.C., 9 December 1994. It is also attributed to Leo Szilard, *The Voice of the Dolphins* (New York: Simon and Schuster, 1961) by Andrew Feenberg in his book *Alternative Modernity: The Technical Turn in Philosophy and Social Theory* (Berkeley, Calif.:University of California Press, 1995), 48.

4. Bush, *Science: The Endless Frontier*.

5. Little has yet been published in English on these developments. In addition to the works cited in chapter 4, n. 7, at least one other book focusing on the "consensus conference" approach is available at this writing: Simon Joss and John Durant, eds., *Public Participation in Science: The Role of Consensus Conferences in Europe* (Cambridge, England: Cambridge University Press, 1995).

6. By contrast, the U.S. Congressional Office of Technology Assessment was abolished in a budget-cutting move by Congress in 1995. See Bruce Bimber, *The Politics of Expertise in Congress: The Rise and Fall of the Office of Technology Assessment* (Albany, N.Y.: State University of New York Press, 1996).

7. "Science shops," sometimes run from storefronts by universities, offer research services, often for free, to ordinary citizens. Richard Sclove of the Loka Institute (Amherst, Massachusetts) is currently coordinating efforts to start a set of science shops in the United States.

8. Martin Bauer, ed., *Resistance to New Technology: Nuclear Power, Information Technology, and Biotechnology* (Cambridge, England: Cambridge University Press, 1995).

9. Shifts of this sort might be expected especially if we have moved into an era in which the simple material output of technology is no longer sufficient to overwhelm any objection or to "buy off" differences among people at large. In this last connection, see Lewis Mumford's discussion of the "enormous bribe" in "Authoritarian and Democratic Technics."

10. "Here Comes the Sun."

11. Philip Weiss, "Off the Grid," *New York Times Magazine*, 8 January 1995, 24–33, 38, 44, 48–52.

12. All of my contacts with mail-order dealers (who are in very close personal touch with their customers by phone) and with installers throughout the country indicate that contacts with militant survivalists are rare. Individuals inclined toward "militia" militancy are, perhaps, less interested in "small is beautiful," low-power approaches than in large generators and more resource-intensive material independence. Remaining survivalists oriented toward very low resource use are very few in number at least as they come into contact with others selling, installing, or using home power systems. (Those who are aware of, but have not spoken directly with, some of the more remotely located home

power adopters may be prone to misinterpret the nature of the desire for independence.)

13. In fact, the linking of home power with "off grid" militancy has been noted within the movement and efforts have been made to counter the misapprehension. A short article referring to the *New York Times Magazine* piece (Weiss, "Off the Grid") and reference to it in *Atlantic Monthly* was published in *Real Goods News* ([Ukiah, California: Real Goods Trading Corp.] summer solstice issue 1995, p. 39) along with a request that readers write to *Atlantic Monthly* and others to correct their misapprehensions. Fortunately, the picture of media coverage is not entirely bad. Occasionally it is far closer to the mark; see, for example, Andi Rierden, "Homeowners Revive Interest in Solar Power," *New York Times*, 25 August 1991, sec. 12, p. 1.

14. David Orr offers one collection of arguments along this line in the specific context of energy decision making. See David Orr, "U.S. Energy Policy and the Political Economy of Participation," *Journal of Politics* 41, no. 4 (November 1979): 1028–56.

15. Interestingly, this seemingly old-fashioned sentiment is widely reflected in engineering and other professional codes of ethics.

16. One of the closest models for such a process may rest in the practices of the Highlander Research and Education Center in New Market, Tennessee, and in the thinking and practice of its founder, Myles Horton.

17. The concealment of the mechanism of technological devices is insightfully handled in Borgmann, *Technology and the Character of Contemporary Life.* See also, Jesse S. Tatum, "Venturing Out into the Open: The Reform of Technology," in *Philosophy in the Service of Things: Devices, Focal Practices, and the Quality of Our Lives,* ed. E. Higgs, A. Light, and D. Strong (Chicago: University of Chicago Press, forthcoming).

18. These ideas were presented in Jesse S. Tatum, "Listening Carefully: Barriers to Democracy in Large Scale Technology" paper presented as part of a panel on "Technoscience, Democracy, and Scale" at the annual meeting of the Society for Social Studies of Science, Charlottesville, Virginia, 18–22 October 1995.

19. Zimmerman, "Toward a More Democratic Ethic of Technological Governance," 86–107 (see 98–99).

20. Winner, *The Whale and the Reactor,* 12.

CHAPTER 10: A BRIEF EPILOGUE

1. I am returning, here, to the quotation used in the introduction: Barber, *Strong Democracy,* 175.

Index